Professor Stewart's Cabinet of Mathematical Curiosities

By the Same Author

Concepts of Modern Mathematics
Game, Set, and Math
Does God Play Dice?
Another Fine Math You've Got Me Into
Fearful Symmetry (with Martin Golubitsky)
Nature's Numbers
From Here to Infinity
The Magical Maze
Life's Other Secret
Flatterland
What Shape is a Snowflake?
The Annotated Flatland (with Edwin A. Abbott)
Math Hysteria
The Mayor of Uglyville's Dilemma
Letters to a Young Mathematician
How to Cut a Cake
Why Beauty is Truth
Taming the Infinite

with Jack Cohen
The Collapse of Chaos
Figments of Reality
What Does a Martian Look Like?
Wheelers (science fiction)
Heaven (science fiction)

with Terry Pratchett and Jack Cohen
The Science of Discworld
The Science of Discworld II: The Globe
The Science of Discworld III: Darwin's Watch

Professor Stewart's Cabinet of Mathematical Curiosities

Ian Stewart

PROFILE BOOKS

First published in Great Britain in 2008 by
PROFILE BOOKS LTD
3a Exmouth House
Pine Street
London EC1R 0JH
www.profilebooks.com

10

Text design by Sue Lamble
Typeset in Stone Serif by Data Standards Ltd, Frome, Somerset.
Printed and bound in Britain by Clays, Bungay, Suffolk.

The moral right of the author has been asserted.

A CIP catalogue record for this book is available from the British Library.

ISBN 978 1 84668 064 9

The paper this book is printed on is certified by the © 1996 Forest
Stewardship Council A.C. CFSC). It is ancient-forest friendly. The printer
holds FSC chain of custody SGS-COC-2061.

FSC

Mixed Sources

Product group from well-managed
forests and other controlled sources

Cert no. SGS-COC-2061
www.fsc.org
© 1996 Forest Stewardship Council

Contents

. . . .

Start Here

● ●

● ● ● There are three kinds of people
      in the world:
      those who can count,
      and those who can't.

When I was fourteen years old, I started a notebook. A maths
notebook. Before you write me off as a sad case, I hasten to add
that it wasn't a notebook of school maths. It was a notebook of
every interesting thing I could find about the maths that *wasn't*
taught at school. Which, I discovered, was a lot, because I soon
had to buy another notebook.

OK, *now* you can write me off. But before you do, have you
spotted the messages in this sad little tale? *The maths you did at
school is not all of it.* Better still: *the maths you* didn't *do at school is
interesting.* In fact, a lot of it is fun – especially when you don't
have to pass a test or get the sums right.

My notebook grew to a set of six, which I still have – and
then spilled over into a filing cabinet when I discovered the
virtues of the photocopier. *Curiosities* is a sample from my
cabinet, a miscellany of intriguing mathematical games, puzzles,
stories and factoids. Most items stand by themselves, so you can
dip in at almost any point. A few form short mini-series. I incline
to the view that a miscellany should be miscellaneous, and this
one is.

The games and puzzles include some old favourites, which
tend to reappear from time to time, and often cause renewed
excitement when they do – the car and the goats, and the twelve-

ball weighing puzzle, both caused a huge stir in the media: one in the USA, the other in the UK. A lot of the material here is new, specially designed for this book. I've striven for variety, so there are logic puzzles, geometrical puzzles, numerical puzzles, probability puzzles, odd items of mathematical culture, things to do and things to make.

One of the virtues of knowing a bit of maths is that you can impress the hell out of your friends. (Be modest about it, though, that's my advice. You can also annoy the hell out of your friends.) A good way to achieve this desirable goal is to be up to speed with the latest buzzwords. So I've scattered some short 'essays' here and there, written in an informal, non-technical style. The essays explain some of the recent breakthroughs that have featured prominently in the media. Things like Fermat's Last Theorem – remember the TV programme? And the four-colour theorem, the Poincaré Conjecture, chaos theory, fractals, complexity science, Penrose patterns. Oh, and there are also some unsolved questions, just to show that maths isn't all *done*. Some are recreational, some serious – like the P = NP? problem, for which a million-dollar prize is on offer. You may not have heard of the problem, but you need to know about the prize.

Shorter, snappy sections reveal interesting facts and discoveries about familiar but fascinating topics: π, prime numbers, Pythagoras's Theorem, permutations, tilings. Amusing anecdotes about famous mathematicians add a historical dimension and give us all a chance to chuckle sympathetically at their endearing foibles.

Now, I did say you could dip in anywhere – and you can, believe me – but to be brutally honest, it's probably better to start at the beginning and dip in following much the same order as the pages. A few of the early items help with later ones, you see. And the early ones tend to be a bit easier, while some of the later ones are, well, a bit ... *challenging*. I've made sure that a lot of easy stuff is mixed in everywhere, though, so that you don't wear your brain out too quickly.

What I'm trying to do is to excite your imagination by showing you lots of amusing and intriguing pieces of mathematics. I want you to have fun, but I'd also be overjoyed if *Curiosities* encouraged you to *engage* with mathematics, experience the thrill of discovery, and keep yourself informed about important developments – be they from four thousand years ago, last week – or tomorrow.

Ian Stewart
Coventry, January 2008

Alien Encounter

The starship *Indefensible* was in orbit around the planet
Noncomposmentis, and Captain Quirk and Mr Crock had
beamed down to the surface.

'According to the *Good Galaxy Guide*, there are two species of
intelligent aliens on this planet,' said Quirk.

'Correct, Captain – Veracitors and Gibberish. They all speak
Galaxic, and they can be distinguished by how they answer
questions. The Veracitors always reply truthfully, and the
Gibberish always lie.'

'But physically—'

'—they are indistinguishable, Captain.'

Quirk heard a sound, and turned to find three aliens creeping
up on them. They looked identical.

'Welcome to Noncomposmentis,' said one of the aliens.

'I thank you. My name is Quirk. Now, you are ...' Quirk
paused. 'No point in asking their names,' he muttered. 'For all we
know, they'll be wrong.'

'That is logical, Captain,' said Crock.

'Because we are poor speakers of Galaxic,' Quirk improvised,
'I hope you will not mind if I call you Alfy, Betty and Gemma.' As
he spoke, he pointed to each of them in turn. Then he turned to
Crock and whispered, 'Not that we know what sex they are,
either.'

'They are all hermandrofemigynes,' said Crock.

'Whatever. Now, Alfy: to which species does Betty belong?'

'Gibberish.'

'Ah. Betty: do Alfy and Gemma belong to different species?'

'No.'

'Right ... Talkative lot, aren't they? Um ... Gemma: to
which species does Betty belong?'

'Veracitor.'

Quirk nodded knowledgeably. 'Right, that's settled it, then!'

'Settled what, Captain?'

'Which species each belongs to.'

'I see. And those species are—?'

'Haven't the foggiest idea, Crock. *You're* the one who's supposed to be logical!'

Answer on page 252

• •

Tap-an-Animal

This is a great mathematical party trick for children. They take turns to choose an animal. Then they spell out its name while you, or another child, tap successive points of the ten-pointed star. You must start at the point labelled 'Rhinoceros', and move in a clockwise direction along the lines. Miraculously, as they say the final letter, you tap the correct animal.

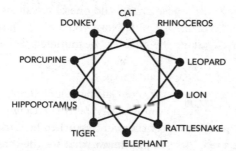

Spell the name to find the animal.

How does it work? Well, the third word along the star is 'Cat', which has three letters, the fourth is 'Lion', with four, and so on. To help conceal the trick, the animals in positions 0, 1 and 2 have 10, 11 and 12 letters. Since 10 taps takes you back to where you started, everything works out perfectly.

To conceal the trick, use *pictures* of the animals – in the diagram I've used their names for clarity.

• •

Curious Calculations

Your calculator can do tricks.

(1) Try these multiplications. What do you notice?

$$1 \times 1$$
$$11 \times 11$$
$$111 \times 111$$
$$1111 \times 1111$$
$$11,111 \times 11,111$$

Does the pattern continue if you use longer strings of 1's?

(2) Enter the number

142,857

(preferably into the memory) and multiply it by 2, 3, 4, 5, 6 and 7. What do you notice?

Answers on page 252

· ·

Triangle of Cards

I have 15 cards, numbered consecutively from 1 to 15. I want to lay them out in a triangle. I've put numbers on the top three for later reference:

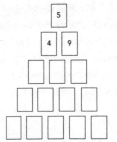

Triangle of cards.

However, I don't want any old arrangement. I want each card to be the difference between the two cards immediately below it, to left and right. For example, 5 is the difference between 4 and 9. (The differences are always calculated so that they are positive.) This condition does not apply to the cards in the bottom row, you appreciate.

The top three cards are already in place – and correct. Can you find how to place the remaining twelve cards?

Mathematicians have found 'difference triangles' like this with two, three or four rows of cards, bearing consecutive whole numbers starting from 1. It has been proved that no difference triangle can have six or more rows.

Answer on page 253

• •

Pop-up Dodecahedron

The *dodecahedron* is a solid made from twelve pentagons, and is one of the five regular solids (page 174).

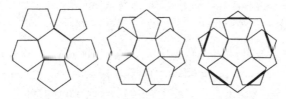

Three stages in making a pop-up dodecahedron.

Cut out two identical copies of the left-hand diagram – 10 cm across – from thickish card. Crease heavily along the joins so that the five pentagonal flaps are nice and floppy. Place one copy on top of the other, like the centre diagram. Lace an elastic band alternately over and under, as in the right-hand diagram (thick solid lines show where the band is on top) – while holding the pieces down with your finger.

Now let go.

If you've got the right size and strength of elastic band, the whole thing will pop up to form a three-dimensional dodecahedron.

Popped-up
dodecahedron.

Sliced Fingers

Here's how to wrap a loop of string around somebody's fingers –
your own or those of a 'volunteer' – so that when it is pulled tight
it seems to slice through the fingers. The trick is striking because
we know from experience that if the string is genuinely linked
with the fingers then it shouldn't slip off. More precisely,
imagine that your fingers all touch a fixed surface – thereby
preventing the string from sliding off their tips. The trick is
equivalent to removing the loop from the holes created by your
fingers and the surface. If the loop were really linked through
those holes, you couldn't remove it at all, so it has to appear to be
linked without actually being linked.

If by mistake it *is* linked, it really would have to slice through
your fingers, so be careful.

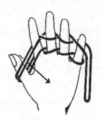

How (not) to slice
your fingers off.

Why is this a mathematical trick? The connection is
topology, a branch of mathematics that emerged over the past
150 years, and is now central to the subject. Topology is about
properties such as being knotted or linked – geometrical features
that survive fairly drastic transformations. Knots remain knotted
even if the string is bent or stretched, for instance.

Make a loop from a 1-metre length of string. Hook one end over the little finger of the left hand, twist, loop it over the next finger, twist in the same direction, and keep going until it passes behind the thumb (left picture). Now bring it round in front of the thumb, and twist it over the fingers in reverse order (right picture). Make sure that when coming back, all the twists are in the *opposite* direction to what they were the first time.

Fold the thumb down to the palm of the hand, releasing the string. Pull hard on the free loop hanging from the little finger … and you can *hear* it slice through those fingers. Yet, miraculously, no damage is done.

Unless you get a twist in the wrong direction somewhere.

Turnip for the Books

'It's been a good year for turnips,' farmer Hogswill remarked to his neighbour, Farmer Suticle.

'Yup, that it has,' the other replied. 'How many did you grow?'

'Well … I don't exactly recall, but I do remember that when I took the turnips to market, I sold six-sevenths of them, plus one-seventh of a turnip, in the first hour.'

'Must've been tricky cuttin' 'em up.'

'No, it was a whole number that I sold. I never cuts 'em.'

'If'n you say so, Hogswill. Then what?'

'I sold six-sevenths of what was left, plus one-seventh of a turnip, in the second hour. Then I sold six-sevenths of what was left, plus one-seventh of a turnip, in the third hour. And finally I sold six-sevenths of what was left, plus one-seventh of a turnip, in the fourth hour. Then I went home.'

'Why?'

''Cos I'd sold the lot.'

How many turnips did Hogswill take to market?

Answer on page 253

The Four-Colour Theorem

Problems that are easy to state can sometimes be very hard to answer. The four-colour theorem is a notorious example. It all began in 1852 with Francis Guthrie, a graduate student at University College, London. Guthrie wrote a letter to his younger brother Frederick, containing what he thought would be a simple little puzzle. He had been trying to colour a map of the English counties, and had discovered that he could do it using four colours, so that no two adjacent counties were the same colour. He wondered whether this fact was special to the map of England, or more general. 'Can every map drawn on the plane be coloured with four (or fewer) colours so that no two regions having a common border have the same colour?' he wrote.

It took 124 years to answer him, and even now, the answer relies on extensive computer assistance. No simple conceptual proof of the four-colour theorem – one that can be checked step by step by a human being in less than a lifetime – is known.

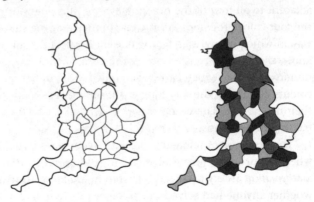

Colouring England's counties with four colours – one solution out of many.

Frederick Guthrie couldn't answer his brother's question, but he 'knew a man who could' – the famous mathematician Augustus De Morgan. However, it quickly transpired that De Morgan *couldn't*, as he confessed in October of the same year in a letter to his even more famous Irish colleague, Sir William Rowan Hamilton.

It is easy to prove that *at least* four colours are necessary for some maps, because there are maps with four regions, each adjacent to all the others. Four counties in the map of England (shown here slightly simplified) form such an arrangement, which proves that at least four colours are necessary in this case. Can you find them on the map?

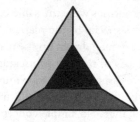

A simple map needing four colours.

De Morgan did make some progress: he proved that it is not possible to find an analogous map with *five* regions, each adjacent to all four of the others. However, this does not prove the four-colour theorem. All it does is prove that the simplest way in which it might go wrong doesn't happen. For all we know, there might be a very complicated map with, say, a hundred regions, which can't be coloured using only four colours because of the way long chains of regions connect to their neighbours. There's no reason to suppose that a 'bad' map has only five regions.

The first printed reference to the problem dates from 1878, when Arthur Cayley wrote a letter to the *Proceedings of the London Mathematical Society* (a society founded by De Morgan) to ask whether anyone had solved the problem yet. They had not, but in the following year Arthur Kempe, a barrister, published a proof, and that seemed to be that.

Kempe's proof was clever. First he proved that any map contains at least one region with five or fewer neighbours. If a region has three neighbours, you can shrink it away, getting a simpler map, and if the simpler map can be 4-coloured, so can the original one. You just give the region that you shrunk whichever colour differs from those of its three neighbours. Kempe had a more elaborate method for getting rid of a region with four or five neighbours. Having established this key fact, the rest of the proof was straightforward: to 4-colour a map, keep shrinking it, region by region, until it has four regions or fewer. Colour those regions with different colours, and then reverse the procedure, restoring regions one by one and colouring them according to Kempe's rules. Easy!

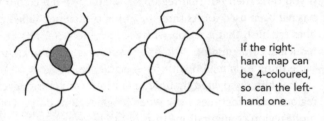

If the right-hand map can be 4-coloured, so can the left-hand one.

It looked too good to be true – and it was. In 1890 Percy Heawood discovered that Kempe's rules didn't always work. If you shrunk a region with five neighbours, and then tried to put it back, you could run into terminal trouble. In 1891 Peter Guthrie Tait thought he had fixed this error, but Julius Petersen found a mistake in Tait's method, too.

Heawood did observe that Kempe's method can be adapted to prove that five colours are always sufficient for any map. But no one could find a map that *needed* more than four. The gap was tantalising, and quickly became a disgrace. When you know that a mathematical problem has either 4 or 5 as its answer, surely you ought to be able to decide which!

But ... no one could.

The usual kind of partial progress then took place. In 1922 Philip Franklin proved that all maps with 26 or fewer regions can be 4-coloured. This wasn't terribly edifying in itself, but Franklin's method paved the way for the eventual solution by introducing the idea of a *reducible configuration*. A configuration is any connected set of regions within the map, plus some information about how many regions are adjacent to those in the configuration. Given some configuration, you can remove it from the map to get a simpler map – one with fewer regions. The configuration is reducible if there is a way to 4-colour the original map, provided you can 4-colour the simpler map. In effect, there has to be a way to 'fill in' colours in that configuration, once everything else has been 4-coloured.

A single region with only three neighbours forms a reducible configuration, for instance. Remove it, and 4-colour what's left – if you can. Then put that region back, and give it a colour that has *not* been used for its three neighbours. Kempe's failed proof does establish that a region with four neighbours forms a reducible configuration. Where he went wrong was to claim the same thing for a region with five neighbours.

Franklin discovered that configurations containing several regions can sometimes work when single regions don't. Lots of multi-region configurations turn out to be reducible.

Kempe's proof would have worked if every region with five neighbours were reducible, and the reason why it would have worked is instructive. Basically, Kempe thought he had proved two things. First, every map contains a region with either three, four or five adjacent ones. Second, each of the associated configurations is reducible. Now these two facts together imply that *every* map contains a reducible configuration. In particular, when you remove a reducible configuration, the resulting simpler map also contains a reducible configuration. Remove that one, and the same thing happens. So, step by step, you can get rid of reducible configurations until the result is so simple that it has at most four regions. Colour those however you wish – at most four colours will be needed. Then restore the previously removed configuration; since this was reducible, the resulting map can also be 4-coloured ... and so on. Working backwards, we eventually 4-colour the original map.

This argument works because *every* map contains one of our irreducible configurations: they form an 'unavoidable set'.

Kempe's attempted proof failed because one of his configurations, a region with five neighbours, isn't reducible. But the message from Franklin's investigation is: don't worry. Try a bigger list, using lots of more complicated configurations. Dump the region with five neighbours; replace it by several configurations with two or three regions. Make the list as big as you need. If you can find *some* unavoidable set of reducible configurations, however big and messy, you're done.

In fact – and this matters in the final proof – you can get away with a weaker notion of unavoidability, applying only to 'minimal criminals': hypothetical maps that require five colours, with the nice feature that any smaller map needs only four colours. This condition makes it easier to prove that a given set is unavoidable. Ironically, once you prove the theorem, it turns out that no minimal criminals exist. No matter: that's the proof strategy.

In 1950 Heinrich Heesch, who had invented a clever method for proving that many configurations are reducible, said that he believed the four-colour theorem could be proved by finding an unavoidable set of reducible configurations. The only difficulty was to find one – and it wouldn't be easy, because some rule-of-thumb calculations suggested that such a set would have to include about 10,000 configurations.

By 1970 Wolfgang Haken had found some improvements to Heesch's method for proving configurations to be reducible, and began to feel that a computer-assisted proof was within reach. It should be possible to write a computer program to check that each configuration in some proposed set is reducible. You could write down several thousand configurations by hand, if you really had to. Proving them unavoidable would be time-consuming, but not necessarily out of reach. But with the computers then available, it would have taken about a century to deal with an unavoidable set of 10,000 configurations. Modern computers can do the job in a few hours, but Haken had to work with what was available, which meant that he had to improve the theoretical methods, and cut the calculation down to a feasible size.

Working with Kenneth Appel, Haken began a 'dialogue' with his computer. He would think of potential new methods for attacking the problem; the computer would then do lots of sums designed to tell him whether these methods were likely to succeed. By 1975, the size of an unavoidable set was down to only 2,000, and the two mathematicians had found much faster tests for irreducibility. Now there was a serious prospect that a human–machine collaboration could do the trick. In 1976 Appel and Haken embarked on the final phase: working out a suitable unavoidable set. They would tell the computer what set they had in mind, and it would then test each configuration to see whether it was reducible. If a configuration failed this test, it was removed and replaced by one or more alternatives, and the computer would repeat the test for irreducibility. It was a delicate process, and there was no guarantee that it would stop – but if it ever did, they would have found an unavoidable set of irreducible configurations.

In June 1976 the process stopped. The computer reported that the current set of configurations – which at that stage contained 1,936 of them, a figure they later reduced to 1,405 – was unavoidable, and every single one of those 1,936 configurations was irreducible. The proof was complete.

The computation took about 1,000 hours in those days, and the test for reducibility involved 487 different rules. Today, with faster computers, we can repeat the whole thing in about an hour. Other mathematicians have found smaller unavoidable sets and improved the tests for reducibility. But no one has yet managed to cut down the unavoidable set to something so small that an unaided human can verify that it does the job. And even if somebody did do that, this type of proof doesn't provide a very satisfactory explanation of why the theorem is true. It just says 'do a lot of sums, and the end result works'. The sums are clever, and there are some neat ideas involved, but most mathematicians would like to get a bit more insight into what's really going on. One possible approach is to invent some notion of 'curvature' for maps, and interpret reducibility as a kind of 'flattening out' process. But no one has yet found a suitable way to do this.

Nevertheless, we now know that the four-colour theorem is true, answering Francis Guthrie's innocent-looking question. Which is an amazing achievement, even if it does depend on a little bit of help from a computer.

Answer on page 254

Answer on page 254

• •

Shaggy Dog Story

Brave Sir Lunchalot was travelling through foreign parts. Suddenly there was a flash of lighting and a deafening crack of thunder, and the rain started bucketing down. Fearing rust, he headed for the nearest shelter, Duke Ethelfred's castle. He arrived to find the Duke's wife, Lady Gingerbere, weeping piteously.

Sir Lunchalot liked attractive young ladies, and for a brief moment he noticed a distinct glint through Gingerbere's tears. Ethelfred was very old and frail, he observed ... Only one thing, he vowed, would deter him from a secret tryst with the Lady – the one thing in all the world that he could not stand.

Puns.

Having greeted the Duke, Lunchalot enquired why Gingerbere was so sad.

'It is my uncle, Lord Elpus,' she explained. 'He died yester-day.'

'Permit me to offer my sincerest condolences,' said Lunchalot.

'That is not why I weep so ... so piteously, sir knight,' replied Gingerbere. 'My cousins Gord, Evan and Liddell are unable to fulfil the terms of uncle's will.'

'Why ever not?'

'It seems that Lord Elpus invested the entire family fortune in a rare breed of giant riding-dogs. He owned 17 of them.'

Lunchalot had never heard of a riding-dog, but he did not

wish to display his ignorance in front of such a lithesome lady. But this fear, it appeared, could be set to rest, for she said, 'Although I have heard much of these animals, I myself have never set eyes on one.'

'They are no fit sight for a fair lady,' said Ethelfred firmly.

'And the terms of the will—?' Lunchalot asked, to divert the direction of the conversation.

'Ah. Lord Elpus left everything to his three sons. He decreed that Gord should receive half the dogs, Evan should receive one-third, and Liddell one-ninth.'

'Mmm. Could be messy.'

'No dog is to be subdivided, good knight.'

Lunchalot stiffened at the phrase 'good knight', but decided it had been uttered innocently and was not a pathetic attempt at humour.

'Well—' Lunchalot began.

'Pah, 'tis a puzzle as ancient as yonder hills!' said Ethelfred scathingly. 'All you have to do is take one of our own riding-dogs over to the castle. Then there are 18 of the damn' things!'

'Yes, my husband, I understand the numerology, but—'

'So the first son gets half that, which is 9; the second gets one-third, which is 6; the third son gets one-ninth, which is 2. That makes 17 altogether, and our own dog can be ridden back here!'

'Yes, my husband, but we have no one here who is manly enough to ride such a dog.'

Sir Lunchalot seized his opportunity. 'Sire, *I* will ride your dog!' The look of admiration in Gingerbere's eye showed him how shrewd his gallant gesture had been.

'Very well,' said Ethelfred. 'I will summon my houndsman and he will bring the animal to the courtyard. Where we shall meet them.'

They waited in an archway as the rain continued to fall. When the dog was led into the courtyard, Lunchalot's jaw dropped so far that it was a good job he had his helmet on. The animal was twice the size of an elephant, with thick striped fur, claws like broadswords, blazing red eyes the size of Lunchalot's shield, huge floppy ears dangling to the ground, and a tail like a pig's – only with more twists and covered in sharp spines. Rain cascaded off its coat in waterfalls. The smell was indescribable.

Perched improbably on its back was a saddle.

Gingerbere seemed even more shocked than he by the sight of this terrible monstrosity. However, Sir Lunchalot was undaunted. *Nothing* could daunt his confidence. *Nothing* could prevent a secret tryst with the Lady, once he returned astride the giant hound, the will executed in full. Except ...

Well, as it happened, Sir Lunchalot did *not* ride the monstrous dog to Lord Elpus's castle, and for all he knows the will has still not been executed. Instead, he leaped on his horse and rode off angrily into the stormy darkness, mortally offended, leaving Gingerbere to suffer the pangs of unrequited lust.

It wasn't Ethelfred's dodgy arithmetic – it was what the Lady had said to her husband in a stage whisper.

What did she say?

Answer on page 254

. . . .

Shaggy Cat Story

No cat has eight tails.
One cat has one tail.
Adding: one cat has nine tails.

. . . .

Rabbits in the Hat

The Great Whodunni, a stage magician, placed his top hat on the table.

'In this hat are two rabbits,' he announced. 'Each of them is either black or white, with equal probability. I am now going to convince you, with the aid of my lovely assistant Grumpelina, that I can deduce their colours without looking inside the hat!'

He turned to his assistant, and extracted a black rabbit from her costume. 'Please place this rabbit in the hat.' She did.

Pop him in the hat and deduce what's already there.

Whodunni now turned to the audience. 'Before Grumpelina added the third rabbit, there were four equally likely combinations of rabbits.' He wrote a list on a small blackboard: BB, BW, WB and WW. 'Each combination is equally likely – the probability is $\frac{1}{4}$.

'But then I added a black rabbit. So the possibilities are BBB, BWB, BBW and BWW – again, each with probability $\frac{1}{4}$.

'Suppose – I won't do it, this is hypothetical – *suppose* I were to pull a rabbit from the hat. What is the probability that it is black? If the rabbits are BBB, that probability is 1. If BWB or BBW, it is $\frac{2}{3}$. If BWW, it is $\frac{1}{3}$. So the overall probability of pulling out a black rabbit is

$$\frac{1}{4} \times 1 + \frac{1}{4} \times \frac{2}{3} + \frac{1}{4} \times \frac{2}{3} + \frac{1}{4} \times \frac{1}{3}$$

which is exactly $\frac{2}{3}$.

'*But* – if there are three rabbits in a hat, of which exactly r are black and the rest white, the probability of extracting a black rabbit is $r/3$. Therefore $r = 2$, so there are two black rabbits in the hat.' He reached into the hat and pulled out a black rabbit. 'Since I added this black rabbit, the original pair must have been one black and one white!'

The Great Whodunni bowed, to tumultuous applause. Then

he pulled two rabbits from the hat – one pale lilac and the other shocking pink.

It seems evident that you can't deduce the contents of a hat without finding out what's inside. Adding the extra rabbit and then removing it again (was it the *same* black rabbit? Do we care?) is a clever piece of misdirection. *But why is the calculation wrong?*

Answer on page 255

. . . .

River Crossing 1 – Farm Produce

Alcuin of Northumbria, aka Flaccus Albinus Alcuinus or Ealhwine, was a scholar, a clergyman and a poet. He lived in the eighth century and rose to be a leading figure at the court of the emperor Charlemagne. He included this puzzle in a letter to the emperor, as an example of 'subtlety in Arithmetick, for your enjoyment'. It still has mathematical significance, as I'll eventually explain. It goes like this.

A farmer is taking a wolf, a goat and a basket of cabbages to market, and he comes to a river where there is a small boat. He can fit only one item of the three into the boat with him at any time. He can't leave the wolf with the goat, or the goat with the cabbages, for reasons that should be obvious. Fortunately the wolf detests cabbage. How does the farmer transport all three items across the river?

Answer on page 256

. . . .

More Curious Calculations

The next few calculator curiosities are variations on one basic theme.

(1) Enter a three-digit number – say 471. Repeat it to get 471,471.

Now divide that number by 7, divide the result by 11, and divide the result by 13. Here we get

$$471,471/7 = 67,353$$
$$67,353/11 = 6,123$$
$$6,123/13 = 471$$

which is the number you first thought of.

Try it with other three-digit numbers – you'll find that exactly the same trick works.

Now, mathematics isn't just about noticing curious things – it's also important to find out *why* they happen. Here we can do that by reversing the entire calculation. The reverse of division is multiplication, so – as you can check – the reverse procedure starts with the three-digit result 471, and gives

$$471 \times 13 = 6,123$$
$$6,123 \times 11 = 67,353$$
$$67,353 \times 7 = 471,471$$

Not terribly helpful as it stands … but what this is telling us is that

$$471 \times 13 \times 11 \times 7 = 471,471$$

So it could be a good idea to see what $13 \times 11 \times 7$ is. Get your calculator and work this out. What do you notice? Does it explain the trick?

(2) Another thing mathematicians like to do is 'generalise'. That is, they try to find related ideas that work in similar ways. Suppose we start with a four-digit number, say 4,715. What should we multiply it by to get 47,154,715? Can we achieve that in several stages, multiplying by a series of smaller numbers?

To get started, divide 47,154,715 by 4,715.

(3) If your calculator runs to ten digits (nowadays a lot of them do), what would the corresponding trick be with five-digit numbers?

(4) If your calculator handles numbers with at least 12 digits, go back to a three-digit number, say 471 again. This time, instead of multiplying it by 7, 11 and 13, try multiplying it by 7, then 11, then 13, then 101, then 9,901. What happens? Why?

(5) Think of a three-digit number, such as 128. Now multiply repeatedly by 3, 3, 3, 7, 11, 13 and 37. (Yes, *three* multiplications by 3.) The result is 127,999,872 – nothing special here. So add the number your first thought of: 128. *Now* what do you get?

Answer on page 257

• •

Extracting the Cherry

This puzzle is a golden oldie, with a simple but elusive answer.

The cocktail cherry is inside the glass, which is formed from four matches. Your task is to move at most two of the matches, so that the cherry is then outside the glass. You can turn the glass sideways or upside down if you wish, but the shape must remain the same.

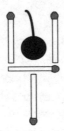

Move two matches to extract the cherry.

Answer on page 258

• •

Make Me a Pentagon

You have a long, thin rectangular strip of paper. Your task is to make from it a regular pentagon (a five-sided figure with all edges the same length and all angles the same size).

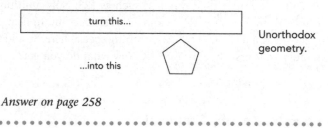

turn this...

...into this

Unorthodox geometry.

Answer on page 258

● ●

What is π?

The number π, which is approximately 3.141 59, is the length of the circumference of a circle whose diameter is exactly 1. More generally, a circle of diameter *d* has a circumference of π*d*. A simple approximation to π is $3\frac{1}{7}$ or 22/7, but this is not exact. $3\frac{1}{7}$ is approximately 3.14 285, which is wrong by the third decimal place. A better approximation is 355/113, which is 3.141 592 9 to seven places, whereas π is 3.141 592 6 to seven places.

How do we know that π is not an exact fraction? However much you continue to improve the approximation *x/y* by using ever larger numbers, you can never get to π itself, only better and better approximations. A number that cannot be written exactly as a fraction is said to be *irrational*. The simplest proof that π is irrational uses calculus, and it was found by Johann Lambert in 1770. Although we can't write down an exact numerical representation of π, we can write down various formulas that define it precisely, and Lambert's proof uses one of those formulas.

More strongly, π is *transcendental* – it does not satisfy any algebraic equation that relates it to rational numbers. This was proved by Ferdinand Lindemann in 1882, also using calculus.

The fact that π is transcendental implies that the classical geometric problem of 'squaring the circle' is impossible. This problem asks for a Euclidean construction of a square whose area is equal to that of a given circle (which turns out to be equivalent to constructing a line whose length is the circumference of the circle). A construction is called Euclidean if it can be performed using an unmarked ruler and a compass. Well, to be pedantic, a 'pair of compasses', which means a single instrument, much as a 'pair of scissors' comes as one gadget.

• •

Legislating the Value of π

There is a persistent myth that the State Legislature of Indiana (some say Iowa, others Idaho) once passed a law declaring the correct value of π to be – well, sometimes people say 3, sometimes $3\frac{1}{6}$ ...

Anyway, the myth is false.

However, something uncomfortably close nearly happened. The actual value concerned is unclear: the document in question seems to imply at least nine different values, all of them wrong. The law was not passed: it was 'indefinitely postponed', and apparently still is. The law concerned was House Bill 246 of the Indiana State Legislature for 1897, and it empowered the State of Indiana to make sole use of a 'new mathematical truth' at no cost. This Bill *was* passed – there was no reason to do otherwise, since it did not oblige the State to do anything. In fact, the vote was unanimous.

The new truth, however, was a rather complicated, and incorrect, attempt to 'square the circle' – that is, to construct π geometrically. An Indianapolis newspaper published an article pointing out that squaring the circle is impossible. By the time the Bill went to the Senate for confirmation, the politicians – even though most of them knew nothing about π – had sensed that there were difficulties. (The efforts of Professor C.A. Waldo of the Indiana Academy of Science, a mathematician who happened to be visiting the House when the Bill was debated, probably helped concentrate their minds.) They did not debate the validity of the mathematics; they decided that the matter was not suitable for legislation. So they postponed the bill ... and as I write, 111 years later, it remains that way.

The mathematics involved was almost certainly the brainchild of Edwin J. Goodwin, a doctor who dabbled in mathematics. He lived in the village of Solitude in Posey County, Indiana, and at various times claimed to have trisected the angle and duplicated the cube – two other famous and equally impossible feats – as well as squaring the circle. At any rate, the legislature of Indiana did not *consciously* attempt to give π an incorrect value by law – although there is a persuasive argument that passing the Bill would have 'enacted' Goodwin's approach, implying its accuracy in law, though perhaps not in mathematics. It's a delicate legal point.

• •

If They *Had* Passed It ...

If the Indiana State Legislature had passed Bill 246, and if the worst-case scenario had proved legally valid, namely that the value of π in law was different from its mathematical value, the consequences would have been distinctly interesting. Suppose that the legal value is $p \neq \pi$, but the legislation states that $p = \pi$. Then

$$\frac{p - \pi}{p - \pi} = 1 \text{ mathematically}$$

but

$$\frac{p - \pi}{p - \pi} = 0 \text{ legally}$$

Since mathematical truths are legally valid, the law would then be maintaining that $1 = 0$. Therefore all murderers have a castiron defence: admit to one murder, then argue that legally it is zero murders. And that's not the last of it. Multiply by one billion, to deduce that one billion equals zero. Now any citizen apprehended in possession of no drugs is in possession of drugs to a street value of $1 billion.

In fact, any statement whatsoever would become legally provable.

It seems likely that the Law would not be *quite* logical enough for this kind of argument to stand up in court. But sillier legal arguments, often based on abuse of statistics, have done just that, causing innocent people to be locked away for long periods. So Indiana's legislators might have opened up Pandora's box.

• •

Empty Glasses

I have five glasses in a row. The first three are full and the other two empty. How can I arrange them so that they are alternately full and empty, by moving only *one* glass?

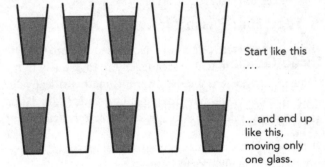

Start like this
...

... and end up like this, moving only one glass.

Answer on page 258

• •

How Many—

Ways are there to rearrange the letters of the (English) alphabet?

403,291,461,126,605,635,584,000,000

Ways are there to shuffle a pack of cards?

80,658,175,170,943,878,571,660,636,856,403,766,975,
289,505,440,883,277,824,000,000,000,000

Different positions are there for a Rubik cube?

43,252,003,274,489,856,000

Different sudoku puzzles are there?

6,670,903,752,021,072,936,960

(Calculated by Bertram Felgenhauer and Frazer Jarvis in 2005.)

Different sequences of 100 zeros and ones are there?

1,267,650,600,228,229,401,496,703,205,376

• •

Three Quickies

(1) After four bridge hands have been dealt, which is the more likely: that you and your partner hold all the spades, or that you and your partner hold no spades?

(2) If you took three bananas from a dish holding thirteen bananas, how many bananas would you have?

(3) A secretary prints out six letters from the computer and addresses six envelopes to their intended recipients. Her boss, in a hurry, interferes and stuffs the letters into the envelopes at random, one letter in each envelope. What is the probability that exactly five letters are in the right envelope?

Answers on page 258

• •

Knight's Tours

The knight in chess has an unusual move. It can jump two squares horizontally or vertically, followed by a single square at right angles, and it hops over any intermediate pieces. The geometry of the knight's move has given rise to many mathematical recreations, of which the simplest is the *knight's tour*. The knight is required to make a series of moves, visiting each square on a chessboard (or any other grid of squares) exactly once. The diagram shows a tour on a 5×5 board, and also shows what the possible moves look like. This tour is not 'closed'– that is, the start and finish squares are not one knight's move apart. Can you find a closed tour on the 5×5 board?

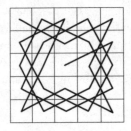 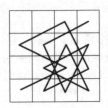

(Left) A 5×5 knight's tour, and (right) a partial 4×4 knight's tour.

I tried to find a 4×4 knight's tour, but I got stuck after visiting 13 squares. Can you find a knight's tour that visits all 16 squares? If not, what is the largest number of squares that the knight can visit?

There is a vast literature on this topic. Good websites include:
www.ktn.freeuk.com
mathworld.wolfram.com/KnightsTour.html

Answers on page 259

• •

Much Undo About Knotting

A mathematician's knot is like an ordinary knot in a piece of string, but the ends of the string are glued together so that the knot can't escape. More precisely, a knot is a closed loop in space. The simplest such loop is a circle, which is called the *unknot*. The next simplest is the *trefoil knot*.

Unknot and trefoil.

Mathematicians consider two knots to be 'the same' – the jargon is *topologically equivalent* – if one can be continuously transformed into the other. 'Continuously' means you have to keep the string in one piece – no cutting – and it can't pass through itself. Knot theory becomes interesting when you discover that a really complicated knot, such as *Haken's Gordian knot*, is in fact just the unknot in disguise.

Haken's Gordian knot.

The trefoil knot is genuine – it can't be unknotted. The first proof of this apparently obvious fact was found in the 1920s.

Knots can be listed according to their complexity, which is measured by the *crossing number* – the number of apparent crossings that occur in a picture of the knot when you draw it using as few crossings as possible. The crossing number of the trefoil knot is 3.

The number of topologically distinct knots with a given number of crossings grows rapidly. Up to 16 crossings, the numbers are:

No. of crossings	3	4	5	6	7	8	9	10
No. of knots	1	1	2	3	7	21	49	165

No. of crossings	11	12	13	14	15	16
No. of knots	552	2176	9,988	46,972	253,293	1,388,705

(For pedants: these numbers refer to *prime knots*, which can't be transformed into two separate knots tied one after the other, and mirror images are ignored.)

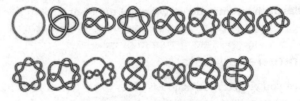

The knots with 7 or fewer crossings.

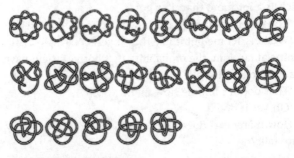

The knots with 8 crossings.

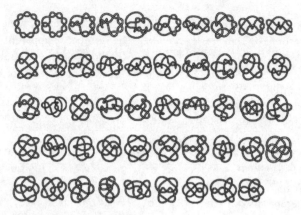

The knots with 9 crossings.

Knot theory is used in molecular biology, to understand knots in DNA, and in quantum physics. Just in case you thought knots were good only for tying up parcels.

For further information see

katlas.math.toronto.edu/wiki/The_Rolfsen_Knot_Table

White-Tailed Cats

'I see you've got a cat,' said Ms Jones to Ms Smith. 'I *do* like its cute white tail! How many cats do you have?'

'Not a lot,' said Ms Smith. 'Ms Brown next door has twenty, which is a lot more than I've got.'

'You still haven't told me how many cats you have!'

'Well ... let me put it like this. If you chose two distinct cats of mine at random, the probability that both of them have white tails is exactly one-half.'

'That doesn't tell me how many you've got!'

'Oh yes it does.'

How many cats does Ms Smith have – and how many have white tails?

Answer on page 259

To Find Fake Coin

In February 2003 Harold Hopwood of Gravesend wrote a short letter to the *Daily Telegraph*, saying that he had solved the newspaper's crossword every day since 1937, but one conundrum had been nagging away at the back of his mind since his schooldays, and at the age of 82 he had finally decided to enlist some help.

The puzzle was this. You are given 12 coins. They all have the same weight, except for one, which may be either lighter or heavier than the rest. You have to find out which coin is different, and whether it is light or heavy, using at most three weighings on a pair of scales. The scales have no graduations for weights; they just have two pans, and you can tell whether they are in balance, or the heavier one has gone down and the lighter one has gone up.

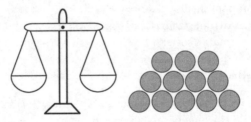

Exactly one coin is either light or heavy: find out which in three weighings.

Before reading on, you should have a go. It's quite addictive.

Within days, the paper's letters desk had received 362 letters and calls about the puzzle, nearly all asking for the answer, and they phoned me. I recognised the problem as one of the classic puzzles, typical of the 'weights and scales' genre, but I'd forgotten the answer. But my friend Marty, who happened to be in the room when I answered the phone, also recognised the problem. The same puzzle had inspired him as a teenager, and his successful solution had led him to become a mathematician.

Of course he had forgotten how the solution went, but we

came up with a method in which we weighed various sets of coins against various others, and faxed it to the newspaper.

In fact, there are many answers, including a very clever one which I finally remembered on the day that the *Telegraph* printed our less elegant method. I had seen it twenty years earlier in *New Scientist* magazine, and it had been reproduced in Thomas H. O'Beirne's *Puzzles and Paradoxes*, which I had on my bookshelf.

Puzzles like this seem to come round every twenty years or so, presumably when a new generation is exposed to them, a bit like an epidemic that gets a new lease of life when the population loses all immunity. O'Beirne traced it back to Howard Grossman in 1945, but it is almost certainly much older, going back to the seventeenth century. It wouldn't surprise me if one day we find it on a Babylonian cuneiform tablet.

O'Beirne offered a 'decision tree' solution, along the lines that Marty and I had concocted. He also recalled the elegant 1950 solution published by 'Blanche Descartes' in *Eureka*, the journal of the Archimedeans, Cambridge University's under-graduate mathematics society. Ms Descartes was in actuality Cedric A.B. Smith, and his solution is a masterpiece of ingenuity. It is presented as a poem about a certain Professor Felix Fiddlesticks, and the main idea goes like this:

> F set the coins out in a row
>> And chalked on each a letter, so,
> To form the words 'F AM NOT LICKED'
>> (An idea in his brain had clicked.)
> And now his mother he'll enjoin:
>> MA DO LIKE
>> ME TO FIND
>> FAKE COIN

This cryptic list of three weighings, one set of four against another, solves the problem, as *Eureka* explains, also in verse. To convince you, I'm going to list all the outcomes of the weighings, according to which coin is heavy or light. Here

L means that the left pan goes down, R that the right pan goes down, and – means they stay balanced.

False coin	1st weighing	2nd weighing	3rd weighing
F heavy	—	R	L
F light	—	L	R
A heavy	L	—	L
A light	R	—	R
M heavy	L	L	—
M light	R	R	—
N heavy	—	R	R
N light	—	L	L
O heavy	L	L	R
O light	R	R	L
T heavy	—	L	—
T light	—	R	—
L heavy	R	—	—
L light	L	—	—
I heavy	R	R	R
I light	L	L	L
C heavy	—	—	R
C light	—	—	L
K heavy	R	—	L
K light	L	—	R
E heavy	R	L	L
E light	L	R	R
D heavy	L	R	—
D light	R	L	—

You can check that no two possibilities give the same results.

The *Telegraph*'s publication of a valid solution did not end the matter. Readers wrote in to object to our answer, on spurious grounds. They wrote to improve it, not always by valid methods. They e-mailed to point out Ms Descartes's solution or similar ones. They told us about other weighing puzzles. They thanked us for setting their minds at rest. They cursed us for reopening an old wound. It was as if some vast, secret reservoir of folk wisdom had suddenly been breached. One correspondent remembered that the puzzle had featured on BBC television in the 1960s, with the solution being given the following night. Ominously, the letter continued, 'I do not recall why it was raised in the first place, or whether that was my first acquaintance with it; *I have a feeling that it was not.*'

• •

Perpetual Calendar

In 1957 John Singleton patented a desk calendar that could represent any date from 01 to 31 using two cubes, but he let the patent lapse in 1965. Each cube bears six digits, one on each face.

A two-cube calendar, and two of the days it can represent.

The picture shows how such a calendar represents the 5th and the 25th day of the month. I have intentionally omitted any other numbers from the faces. You are allowed to place the cubes with any of the six faces showing, and you can also put the grey one on the left and the white one on the right.

What are the numbers on the two cubes?

Answer on page 260

• •

Mathematical Jokes 1*

A biologist, a statistician and a mathematician are sitting outside a cafe watching the world go by. A man and a woman enter a building across the road. Ten minutes later, they come out accompanied by a child.

'They've reproduced,' says the biologist.

* The purpose of these jokes is not primarily to make you laugh. It is to show you what makes *mathematicians* laugh, and to provide you with a glimpse into an obscure corner of the world's mathematical subculture.

'No,' says the statistician. 'It's an observational error. On average, two and a half people went each way.'

'No, no, no,' says the mathematician. 'It's perfectly obvious. If someone goes in now, the building will be empty.'

• •

Deceptive Dice

The Terrible Twins, Innumeratus and Mathophila, were bored.

'I know,' said Mathophila brightly. 'Let's play dice!'

'Don't like dice.'

'Ah, but these are *special* dice,' said Mathophila, digging them out of an old chocolate box. One was red, one yellow and one blue.

Innumeratus picked up the red dice.* 'There's something funny about this one,' he said. 'It's got two 3's, two 4's and two 8's.'

'They're all like that,' said Mathophila carelessly. 'The yellow one has two 1's, two 5's and two 9's – and the blue one has two 2's, two 6's and two 7's.'

'They look rigged to me,' said Innumeratus, deeply suspicious.

'No, they're perfectly fair. Each face has an equal chance of turning up.'

'How do we play, anyway?'

'We each choose a different one. We roll them simultaneously, and the highest number wins. We can play for pocket money.' Innumeratus looked sceptical, so his sister quickly added: 'Just to be fair, I'll let you choose first! Then you can choose the best dice!'

'Weeelll ... ' said Innumeratus, hesitating.

Should he play? If not, why not?

Answer on page 260

• •

* Strictly speaking, 'dice' is the plural, and I should have used 'die' – but I've given up fighting that particular battle. I mention this to stop people writing in to tell me I've got it wrong. Anyway, the proverb tells us 'never say die'.

An Age-Old Old-Age Problem

The Emperor Scrumptius was born in 35 BC, and died on his birthday in AD 35. What was his age when he died?

Answer on page 262

● ●

Why Does Minus Times Minus Make Plus?

When we first meet negative numbers, we are told that multiplying two negative numbers together makes a positive number, so that, for example, $(-2) \times (-3) = +6$. This often seems rather puzzling.

The first point to appreciate is that starting from the usual conventions for arithmetic with positive numbers, we are free to define $(-2) \times (-3)$ to be anything we want. It could be -99, or 127π, if we wished. So the main question is not what is the true value, but what is the sensible value. Several different lines of thought all converge on the same result – namely, that $(-2) \times (-3) = +6$. I include the $+$ sign for emphasis.

But *why* is this sensible? I rather like the interpretation of a negative number as a debt. If my bank account contains £–3, then I owe the bank £3. Suppose that my debt is multiplied by 2 (positive): then it surely becomes a debt of £6. So it makes sense to insist that $(+2) \times (-3) = -6$, and most of us are happy with that. What, though, should $(-2) \times (-3)$ be? Well, if the bank kindly writes off (takes away) two debts of £3 each, I am £6 better off – my account has changed exactly as it would if I had deposited £+6. So in banking terms, we want $(-2) \times (-3)$ to equal $+6$.

The second argument is that we can't have both $(+2) \times (-3)$ and $(-2) \times (-3)$ equal to $+6$. If that were the case, then we could cancel the -3 and deduce that $+2 = -2$, which is silly.

The third argument begins by pointing out an unstated assumption in the second one: that the usual laws of arithmetic should remain valid for negative numbers. It proceeds by adding that this is a reasonable thing to aim for, if only for mathematical elegance. If we require the usual laws to be valid, then

$$(+2) \times (-3) + (-2) \times (-3) = (2-2) \times (-3) = 0 \times (-3) = 0$$

so

$$-6 + (-2) \times (-3) = 0$$

Adding 6 to both sides, we find that

$$(-2) \times (-3) = +6$$

In fact a similar argument justifies taking $(+2) \times (-3) = -6$, as well.

Putting all this together: mathematical elegance leads us to *define* minus times minus to be plus. In applications such as finance, this choice turns out to match reality in a straightforward manner. So as well as keeping arithmetic simple, we end up with a good model for important aspects of the real world.

We could do it differently. But we'd end up by complicating arithmetic, and reducing its applicability. Basically, there's no contest. Even so, 'minus times minus makes plus' is a conscious human convention, not an inescapable fact of nature.

• •

Heron Suit

No cat that wears a heron suit is unsociable.
No cat without a tail will play with a gorilla.
Cats with whiskers always wear heron suits.
No sociable cat has blunt claws.
No cats have tails unless they have whiskers.
Therefore:
No cat with blunt claws will play with a gorilla.
Is the deduction logically correct?

Answer on page 262

• •

How to Unmake a Greek Cross

To paraphrase the old music-hall joke, almost any insult will make a Greek cross. But what I want you to do is *unmake* a Greek cross. In this region of Puzzledom, a Greek cross is five equal squares joined to make a + shape. I want you to convert it to a square, by cutting it into pieces and reassembling them. Here's one solution, using five pieces. But can you find an alternative, using four pieces, *all the same shape*?

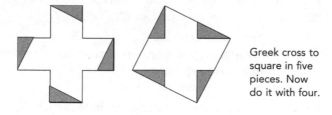

Greek cross to square in five pieces. Now do it with four.

Answer on page 263

How to Remember a Round Number

A traditional French rhyme goes like this:*

> Que j'aime a faire apprendre
> Un nombre utile aux sages!
> Glorieux Archimède, artiste ingenieux,
> Toi, de qui Syracuse loue encore le mérite!

But to which 'number useful to the sages' does it refer? Counting the letters in each word, treating 'j' as a word with one letter and placing a decimal point after the first digit, we get

3.141 592 653 589 793 238 462 6

---

* A loose translation is:
> How I like to make
> The sages learn a useful number!
> Glorious Archimedes, ingenious artist,
> You whose merit Syracuse still praises.

which is π to the first 22 decimal places. Many similar mnemonics for π exist in many languages. In English, one of the best known is

> How I want a drink, alcoholic of course, after the heavy chapters involving quantum mechanics. One is, yes, adequate even enough to induce some fun and pleasure for an instant, miserably brief.

It probably stopped there because the next digit is a 0, and it's not entirely clear how best to represent a word with no letters. (For one convention, see later.) Another is:

> Sir, I bear a rhyme excelling
> In mystic force, and magic spelling
> Celestial sprites elucidate
> All my own striving can't relate.

An ambitious π-mnemonic featured in *The Mathematical Intelligencer* in 1986 (volume 8, page 56). This is an informal 'house journal' for professional mathematicians. The mnemonic is a self-referential story encoding the first 402 decimals of π. It uses punctuation marks (ignoring full stops) to represents the digit zero, and words with more than 9 letters represent two consecutive digits – for instance, a word with 13 letters represents the digits 13 in that order. Oh, and any actual digit represents itself. The story begins like this:

> For a time I stood pondering on circle sizes. The large computer mainframe quietly processed all of its assembly code. Inside my entire hope lay for figuring out an elusive expansion. Value: pi. Decimals expected soon. I nervously entered a format procedure. The mainframe processed the request. Error. I, again entering it, carefully retyped. This iteration gave zero error printouts in all – success.

For the rest of the story, and many other π-mnemonics in various languages, see
www.geocities.com/capecanaveral/lab/3550/pimnem.htm

The Bridges of Königsberg

Occasionally, a simple puzzle starts a whole new area of mathematics. Such occurrences are rare, but I can think of at least three. The most famous such puzzle is known as the Bridges of Königsberg, which led Leonhard Euler* to invent a branch of graph theory in 1735. Königsberg, which was in Prussia in those days, straddled the river Pregel. There were two islands, connected to the banks and each other by seven bridges. The puzzle was: is it possible to find a path that crosses each bridge exactly once?

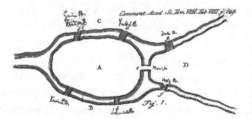

Euler's diagram of the Königsberg bridges.

Euler solved the puzzle by proving that no solution exists. More generally, he gave a criterion for any problem of this kind to have a solution, and observed that it did not apply to this particular example. He pointed out that the exact geometry is irrelevant – what matters is how everything is connected. So the puzzle reduces to a simple network of dots, joined by lines, here shown superimposed on the map. Each dot corresponds to one connected piece of land, and two dots are joined by lines if there is a bridge linking the corresponding pieces of land.

---

* It is mandatory to point out that his name is pronounced 'oiler', not 'yooler'. Numerous oil-based puns then become equally mandatory, but I won't mention any.

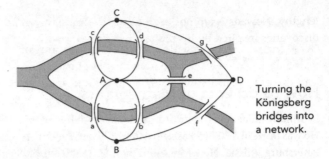

Turning the Königsberg bridges into a network.

So we get four dots, A, B, C and D, and seven edges, a, b, c, d, e, f and g, one for each bridge. The puzzle now simplifies to this: is it possible to find a path through the network that includes each line exactly once? You might like to experiment before reading on.

To work out when a solution exists, Euler distinguished two kinds of path. An *open tour* starts and ends at different dots; a *closed tour* starts and ends at the same dot. He proved that for this particular network, neither kind of tour exists. The main theoretical idea is the *valency* of each dot: how many lines meet there. For instance, 5 lines meet at dot A, so the valency of A is 5.

Suppose that a closed tour exists on some network. Whenever one of the lines in the tour enters a dot, then the next line must exit from that dot. So, if a closed tour is possible, the number of lines at any given dot must be even: every dot must have even valency. This already rules out any closed tour of the Königsberg bridges, because that network has three dots of valency 3 and one of valency 5 – all odd numbers.

A similar criterion works for open tours, but now there must be exactly two dots of odd valency: one at the start of the tour, the other at its end. The Königsberg diagram has four vertices with odd valency, so there is no open tour either.

Euler also proved that these conditions are sufficient for a tour to exist, provided the diagram is connected – any two dots must be linked by *some* path. Euler's proof of this is quite lengthy. Nowadays a proof takes just a few lines, thanks to new discoveries inspired by his pioneering efforts.

● ●

How to do Lots of Mathematics

Leonhard Euler was the most prolific mathematician of all time. He was born in 1707 in Basel, Switzerland, and died in 1783 in St Petersburg, Russia. He wrote more than 800 research papers, and a long list of books. Euler had 13 children, and often worked on his mathematics while one of them sat on his knee. He lost the sight of one eye in 1735, probably because of a cataract; the other eye failed in 1766. Going blind seems to have had no effect on his productivity. His family took notes, and he had an astonishing mental powers – once doing a mental calculation to fifty decimal places to decide which of two students had the right answer.

Leonhard Euler.

Euler spent many years at the court of Queen Catherine the Great. It has been suggested that to avoid becoming embroiled in court politics – which could easily prove fatal – Euler spent nearly all of his time working on mathematics, except when he was asleep. That way it was obvious that he had no time for intrigue.

Which reminds me of a mathematical joke: Why should a mathematician keep a mistress as well as a wife? (For gender-equality reasons, feel free to change to 'a lover as well as a husband'.) Answer: when the wife thinks you're with the mistress, and the mistress thinks you're with the wife, you have time to get on with your mathematics.

. . . .

Euler's Pentagonal Holiday

Here's your chance to put Euler's discoveries about tours on networks to the test. (a) Find an open tour of this network. (b) Find one that looks the same when you reflect the figure to interchange left and right.

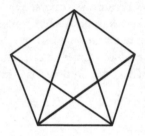

A network with an open tour.

Answer on page 263

. . . .

Ouroborean Rings

Around 1960 the American mathematician Sherman K. Stein discovered a curious pattern in the Sanskrit nonsense word *yamátárájabhánasalagám*. The composer George Perle told Stein that the stressed (*á*) and unstressed (*a*) syllables form a mnemonic for rhythms, and correspond to long and short beats. Thus the first three syllables, *ya má tá*, have the rhythm short, long, long. The second to fourth are *má tá rá*, long, long, long – and so on. There are eight possible triplets of long or short rhythms, and each occurs in the nonsense word exactly once.

Stein rewrote the word using 0 for short and 1 for long,

getting 0111010001. Then he noticed that the first two digits are the same as the last two, so the string of digits can be bent into a loop, swallowing its own tail. Now you can generate all possible sequences of three digits 0 and 1 by moving along the loop one space at a time:

I call such sequences *ouroborean rings*, after the mythical serpent Ouroboros, which eats its own tail.

There is an ouroborean ring for pairs: 0011. It is unique except for rotations. Your task is to find one for quadruplets. That is, arrange eight 0's and eight 1's in a ring so that every possible string of four digits, from 0000 to 1111, appears as a series of consecutive symbols. (Each string of four must then occur exactly once.)

Answer on page 263

The Ourotorus

Are there higher-dimensional analogues of ouroborean rings?

For example, there are sixteen 2×2 squares with entries 0 or 1. Is it possible to write 0's and 1's in a 4×4 square so that each possibility occurs exactly once as a subsquare? You must pretend that opposite edges of the square are joined together, so that it wraps round into an *ourotorus*.

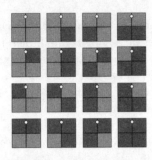

16 pieces for the
ourotorus puzzle.

You can turn this puzzle into a game. Cut out the sixteen
pieces shown – the small dot near the top tells you which way up
they go. Can you arrange them in a 4 × 4 grid, keeping the dot at
the top, so that adjacent squares have the same colours along
common edges? This rule also applies to squares that become
adjacent if the top and bottom, or the left and right sides, of the
grid are 'wrapped round' so that they join.

Answer on page 264

● ●

Who Was Pythagoras?

We recognise the name 'Pythagoras' because it is attached to a
theorem, one that most of us have grappled with at school. 'The
square on the hypotenuse of a right-angled triangle is equal to
the sum of the squares on the other two sides.' That is, if you take
any right-angled triangle, then the square of the longest side is
equal to the sum of the squares of the other two sides. Well
known as his theorem may be, the actual person has proved
rather elusive, although we know more about him as a historical
figure than we do for, say, Euclid. What we *don't* know is whether
he proved his eponymous theorem, and there are good reasons
to suppose that, even if he did, he wasn't the first to do so.

But more of that story later.

Pythagoras was Greek, born around 569 BC on the island of
Samos in the north-eastern Aegean. (The exact date is disputed,
but this one is wrong by at most 20 years.) His father,

Mnesarchus, was a merchant from Tyre; his mother, Pythais, was from Samos. They may have met when Mnesarchus brought corn to Samos during a famine, and was publicly thanked by being made a citizen.

Pythagoras studied philosophy under Pherekydes. He probably visited another philosopher, Thales of Miletus. He attended lectures given by Anaximander, a pupil of Thales, and absorbed many of his ideas on cosmology and geometry. He visited Egypt, was captured by Cambyses II, the King of Persia, and taken to Babylon as a prisoner. There he learned Babylonian mathematics and musical theory. Later he founded the school of Pythagoreans in the Italian city of Croton (now Crotone), and it is for this that he is best remembered. The Pythagoreans were a mystical cult. They believed that the universe is mathematical, and that various symbols and numbers have a deep spiritual meaning.

Various ancient writers attributed various mathematical theorems to the Pythagoreans, and by extension to Pythagoras – notably his famous theorem about right-angled triangles. But we have no idea what mathematics Pythagoras himself originated. We don't know whether the Pythagoreans could prove the theorem, or just believed it to be true. And there is evidence from the inscribed clay tablet known as Plimpton 322 that the ancient Babylonians may have understood the theorem 1200 years earlier – though they probably didn't possess a proof, because Babylonians didn't go much for proofs anyway.

• •

Proofs of Pythagoras

Euclid's method for proving Pythagoras's Theorem is fairly complicated, involving a diagram known to Victorian school-boys as 'Pythagoras's pants' because it looked like underwear hung on a washing line. This particular proof fitted into Euclid's development of geometry, which is why he chose it. But there are many other proofs, some of which make the theorem much more obvious.

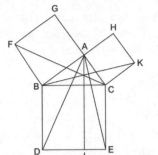

Pythagoras's pants.

One of the simplest proofs is a kind of mathematician's jigsaw puzzle. Take any right-angled triangle, make four copies, and assemble them inside a carefully chosen square. In one arrangement we see the square on the hypotenuse; in the other, we see the squares on the other two sides. Clearly, the areas concerned are equal, since they are the difference between the area of the surrounding square and the areas of the four copies of the triangle.

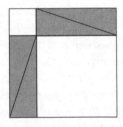

(Left) The square on the hypotenuse (plus four triangles). (Right) The sum of the squares on the other two sides (plus four triangles). Take away the triangles ... and Pythagoras's Theorem is proved.

Then there's a cunning tiling pattern. Here the slanting grid is formed by copies of the square on the hypotenuse, and the other grid involves both of the smaller squares. If you look at how one slanting square overlaps the other two, you can see how to cut the big square into pieces that can be reassembled to make the two smaller squares.

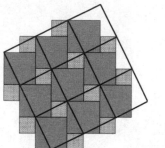

Proof by
tiling.

Another proof is a kind of geometric 'movie', showing how to
split the square on the hypotenuse into two parallelograms,
which then slide apart – without changing area – to make the
two smaller squares.

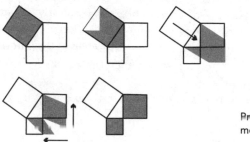

Proof by
movie.

● ●

A Constant Bore

'Now, *this* component is a solid copper sphere with a cylindrical
hole bored exactly through its centre,' said Rusty Nail, the
construction manager. He opened a blueprint on his laptop's
screen:

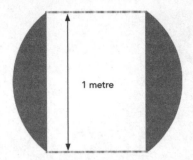

Cross-section of sphere
with cylindrical hole.

'Looks straightforward,' said the foreman, Lewis Bolt. 'That's quite a lot of copper.'

'Coincidentally, that's what I want you to work out,' said Rusty. 'What volume of copper do we need to cast it?'

Lewis stared at the blueprint. 'It doesn't say how big the sphere is.' He paused. 'I can't find the answer unless you tell me the radius of the sphere.'

'Hmmm,' said Rusty. 'They must have forgotten to put that in. But I'm sure you can work something out. I need the answer by lunchtime.'

What is the volume of copper required? Does it depend on the size of the sphere?

Answer on page 264

• •

Fermat's Last Theorem

The great virtue of Fermat's Last Theorem is that it's easy to understand what it means. What made the theorem famous is that proving it turned out to be amazingly hard. So hard, in fact, that it took about 350 years of effort, by many of the world's leading mathematicians, to polish it off. And to do that, they had to invent entire new mathematical theories, and prove things that looked much harder.

Pierre de Fermat.

It all started around 1650, when Pierre de Fermat wrote an enigmatic note in the margin of his copy of Diophantus's book *Arithmetica*: 'of which fact I have found a remarkable proof, 'but this margin is too small to contain it.' Proof of what? Let me back up a bit.

Diophantus was probably Greek, and he lived in ancient Alexandria. Some time around AD 250 he wrote a book about solving algebraic equations – with a slight twist: the solutions were required to be fractions, or better still, whole numbers. Such equations are called *Diophantine equations* to this day. A typical Diophantine problem is: find two squares whose sum is square (using only whole numbers). One possible answer is 9 and 16, which add up to 25. Here 9 is the square of 3, while 16 is the square of 4, and 25 is the square of 5. Another answer is 25 (the square of 5) and 144 (the square of 12), which add up to 169 (the square of 13). These are the tip of an iceberg.

This particular problem is linked to Pythagoras's Theorem, and Diophantus was following a long tradition of looking for *Pythagorean triples* – whole numbers that can form the sides of a right-angled triangle. Diophantus wrote down a general rule for finding all Pythagorean triples. He wasn't the first to discover it, but it belonged very naturally in his book. Now Fermat wasn't a professional mathematician – he never held an academic position. In his day job he was a legal advisor. But his passion was mathematics, especially what we now call *number theory*, the properties of ordinary whole numbers. This area uses the simplest ingredients anywhere in mathematics, but paradoxically it is one of the most difficult areas to make progress in. The simpler the ingredients, the harder it is to make things with them.

Fermat pretty much created number theory. He took over where Diophantus had left off, and by the time he had finished, the subject was virtually unrecognisable. And some time around 1650 – we don't even know the exact date – he must have been thinking about Pythagorean triples, and wondered 'can we do it with *cubes*?'

Just as the square of a number is what you get by multiplying two copies of the same number, the cube is what you get by multiplying three copies. That is, the square of 5, say, is $5 \times 5 = 25$, and the cube of 5 is $5 \times 5 \times 5 = 125$. These are written more compactly as 5^2 and 5^3, respectively. No doubt Fermat tried a few possibilities. Is the sum of the cubes of 1 and 2 a cube, for instance? The cubes here are 1 and 8, so their sum is 9. That's a square, but not a cube: no banana.

He surely noticed that you can get pretty close. The cube of 9 is 729; the cube of 10 is 1,000; their sum is 1,729. That's *very nearly* the cube of 12, which is 1,728. Missed by one! Still no banana.

Like any mathematician, Fermat would have tried bigger numbers, and used any short cuts he could think of. Nothing worked. Eventually he gave up: he hadn't found any solutions, and by now he suspected that there weren't any. Except for the cube of 0 (which is also 0) and any cube whatsoever, which add up to the whatsoever – but we all know that adding zero makes no difference to anything, so that's 'trivial', and he wasn't interested in trivialities.

OK, so cubes don't get us anywhere. What about the next such type of number, fourth powers? You get those by multiplying four copies of the same number, for example $3 \times 3 \times 3 \times 3 = 81$ is the fourth power of 3, written as 3^4. Still no joy. In fact, for fourth powers Fermat found a logical proof that no solutions exist except trivial ones. Very few of Fermat's proofs have survived, and few of them were written down, but we know how this one went, and it's both cunning and correct. It takes some hints from Diophantus's method of finding Pythagorean triples.

Fifth powers? Sixth powers? Still nothing. By now Fermat was ready to make a bold statement: 'To resolve a cube into the sum of two cubes, a fourth power into two fourth powers, or in general any power higher than the second into two powers of the same kind, is impossible.' That is: the only way for two nth powers to add up to an nth power is when n is 2 and we are looking at Pythagorean triples. This is what he wrote in his margin, and it's what caused so much fuss for the next 350 years.

We don't actually have Fermat's copy of the *Arithmetica* with its marginal notes. What we have is a printed edition of the book prepared later by his son, which has the notes printed in it.

Fermat included various other unproved but fascinating bits of number theory in his letters and the marginal notes published by his son, and the world's mathematicians rose to the challenge. Soon all but one of Fermat's statements had been proved – apart from one that was disproved, but in that case Fermat never claimed he had a proof anyway. The sole remaining 'last theorem' – not the last one he wrote down, but the last one that no one else could prove or disprove – was the marginal note about sums of like powers.

Fermat's Last Theorem became notorious. Euler proved that there is no solution in cubes. Fermat himself had done fourth powers. Peter Lejeune Dirichlet dealt with fifth powers in 1828, and 14th powers in 1832. Gabriel Lamé published a proof for 7th powers, but it had a mistake in it. Carl Friedrich Gauss, one of the best mathematicians who has ever lived and an expert in number theory, tried to patch up Lamé's attempt, but failed, and gave up on the whole problem. He wrote to a scientific friend that the problem 'has little interest for me, since a multitude of such propositions, which one can neither prove nor refute, can easily be formulated'. But for once Gauss's instincts let him down: the problem *is* interesting, and his remark seems to have been a case of sour grapes.

In 1874 Lamé had a new idea, linking Fermat's Last Theorem to a special type of complex number – involving the square root of minus one (see page 184). There was nothing wrong with complex numbers, but there was a hidden assumption in Lamé's argument, and Ernst Kummer wrote to him to inform him that it went wrong for 23rd powers. Kummer managed to fix Lamé's idea, eventually proving Fermat's Last Theorem for all powers up to the 100th, except for 37, 59 and 67. Later mathematicians polished off these powers too, and extended the list, until by 1980 Fermat's Last Theorem had been proved for all powers up to the 125,000th.

You might think that this would be good enough, but mathematicians are made of sterner stuff. It has to be *all* powers, or nothing. The first 125,000 whole numbers are minuscule compared with the infinity of numbers that remain. But Kummer's methods needed special arguments for each power, and they weren't really up to the job. What was needed was *a new idea*. Unfortunately, nobody knew where to look for one.

So number theorists abandoned Fermat's Last Theorem and headed off into areas where they could still make progress. One such area, the theory of *elliptic curves*, started to get really exciting, but also very technical. An elliptic curve is not an ellipse – if it were, we wouldn't need a different name for it. It is a curve in the plane whose *y*-coordinate, squared, is a cubic formula in its *x*-coordinate. These curves in turn are connected with some remarkable expressions involving complex numbers, called *elliptic functions*, which were in vogue in the late nineteenth century. The theory of elliptic curves, and their associated elliptic functions, became very deep and powerful.

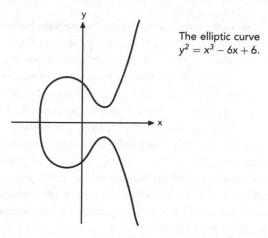

The elliptic curve
$y^2 = x^3 - 6x + 6$.

Starting around 1970, a series of mathematicians started to get glimpses of a strange connection between elliptic curves and Fermat's Last Theorem. Roughly speaking, if Fermat was wrong, and two nth powers do add up to another nth power, then those three numbers determine an elliptic curve. And because the powers add like that, it is a very strange elliptic curve, with a surprising combination of properties. So surprising, in fact, that it looks wildly unlikely that it can exist at all, as Gerhard Frey pointed out in 1985.

This observation opens the way to a 'proof by contradiction', what Euclid called '*reductio ad absurdum*' (reduction to the absurd). To prove that some statement is true, you begin by assuming that, on the contrary, it is false. Then you deduce the logical consequences of this falsity. If the consequences contradict each other or known facts, then your assumption must have been wrong – so the statement must be true after all. In 1986 Kenneth Ribet pinned this idea down by proving that if Fermat's Last Theorem is false, then the associated elliptic curve violates a conjecture (that is, a plausible but unproved theorem) introduced by the Japanese mathematicians Yutaka Taniyama and Goro Shimura. This Taniyama–Shimura conjecture, which dates from 1955, says that every elliptic curve is associated with a special class of elliptic functions, called *modular functions*.

Ribet's discovery implies that any proof of the Taniyama–Shimura conjecture automatically proves – by contradiction – Fermat's Last Theorem as well. Because the assumed falsity of Fermat's Last Theorem tells us that Frey's elliptic curve exists, but the Taniyama–Shimura conjecture tells us that it doesn't.

Unfortunately, the Taniyama–Shimura conjecture was just that – a conjecture.

Enter Andrew Wiles. When Wiles was a child he heard about Fermat's Last Theorem, and decided that when he grew up he would become a mathematician and prove it. He did become a mathematician, but by then he had decided that Fermat's Last Theorem was much as Gauss had complained – an isolated question of no particular interest for the mainstream of mathematics. But Frey's discovery changed all that. It meant that Wiles could work on the Taniyama–Shimura conjecture, an important mainstream problem, and polish off Fermat's Last Theorem too.

Now, the Taniyama–Shimura conjecture is very difficult – that's why it remained a conjecture for some forty years. But it has good links to many areas of mathematics, and sits firmly in the middle of an area where the techniques are very powerful: elliptic curves. For seven years Wiles worked away in his study, trying every technique he could think of, striving to prove the Taniyama–Shimura conjecture. Hardly anybody knew that he was working on that problem; he wanted to keep it secret.

In June 1993 Wiles gave a series of three lectures at the Isaac Newton Institute in Cambridge, one of the world's top mathematical research centres. Their title was 'Modular forms, elliptic curves and Galois representations', but the experts knew it must really be about the Taniyama–Shimura conjecture – and, just possibly, Fermat's Last Theorem. On day three, Wiles announced that he had proved the Taniyama–Shimura conjecture, not for all elliptic curves, but for a special kind called 'semistable'.

Frey's elliptic curve, if it exists, is semistable. Wiles was telling his audience that he had proved Fermat's Last Theorem.

But it wasn't quite that straightforward. In mathematics you don't get credit for solving a big problem by giving a few lectures in which you say you've got the answer. You have to publish your ideas in full, so that everyone else can check that they are right. And when Wiles started that process – which involves getting experts to go over the work in detail before it gets into print – some logical gaps emerged. He quickly filled most of them, but one seemed much harder, and it wouldn't go away. As rumours spread that the proposed proof had collapsed, Wiles made one final attempt to shore up his increasingly rickety proof – and, contrary to most expectations, he succeeded. One final technical point was supplied by his former student, Richard Taylor, and by the end of October 1994 the proof was complete. The rest, as they say, is history.

By developing Wiles's new methods, the Taniyama–Shimura conjecture has now been proved for all elliptic curves, not just semistable ones. And although the *result* of Fermat's Last Theorem is still just a minor curiosity – nothing important rests on it being true or false – the *methods* used to prove it have become a permanent and important addition to the mathematical armoury.

One question remains. Did Fermat really have a valid proof, as he claimed in his margin? If he did, it certainly wasn't the one that Wiles found, because the necessary ideas and methods simply did not exist in Fermat's day. As an analogy: today we could erect the pyramids using huge cranes, but we can be confident that however the ancient Egyptians built their pyramids, they didn't use modern machinery. Not just because there is no evidence of such machines, but because the necessary infrastructure could not have existed. If it had done, the whole culture would have been different. So the consensus among mathematicians is that what Fermat thought was a proof probably had a logical gap that he missed. There are some plausible but incorrect attempts that would have been feasible in his day. But we don't know whether his proof – if one ever existed – followed those lines. Maybe – just maybe – there is a much simpler proof lurking out there in some unexplored realm of mathematical imagination, waiting for somebody to stumble into it.* Stranger things have happened.

• •

Pythagorean Triples

I can't really get away without telling you Diophantus's method for finding all Pythagorean triples, can I?

OK, here it is. Take any two whole numbers, and form:

- twice their product
- the difference between their squares
- the sum of their squares

Then the resulting three numbers are the sides of a Pythagorean triangle.

For instance, take the numbers 2 and 1. Then

- twice their product $= 2 \times 2 \times 1 = 4$
- the difference between their squares $= 2^2 - 1^2 = 3$
- the sum of their squares $= 2^2 + 1^2 = 5$

and we obtain the famous 3–4–5 triangle. If instead we take numbers 3 and 2, then

- twice their product $= 2 \times 3 \times 2 = 12$
- the difference between their squares $= 3^2 - 2^2 = 5$
- the sum of their squares $= 3^2 + 2^2 = 13$

and we get the next-most-famous 5–12–13 triangle. Taking numbers 42 and 23, on the other hand, leads to

- twice their product $= 2 \times 42 \times 23 = 1,932$

---

* If you think you've found it, *please don't send it to me*. I get too many attempted proofs as it is, and so far – well, just don't get me started, OK?

- the difference between their squares $= 42^2 - 23^2 = 1,235$
- the sum of their squares $= 42^2 + 23^2 = 2,293$

and no one has ever heard of the $1,235 - 1,932 - 2,293$ triangle. But these numbers do work:

$$1,235^2 + 1,932^2 = 1,525,225 + 3,732,624 = 5,257,849$$
$$= 2,293^2$$

There's a final twist to Diophantus's rule. Having worked out the three numbers, we can choose any other number we like and multiply them all by that. So the 3–4–5 triangle can be converted to a 9–12–15 triangle by multiplying all three numbers by 3, or to an 18–24–30 triangle by multiplying all three numbers by 6. We can't get these two triples from the above prescription using whole numbers. Diophantus knew that.

● ●

Prime Factoids

Prime numbers are among the most fascinating in the whole of mathematics. Here's a Prime Primer.

A whole number bigger than 1 is *prime* if it is not the product of two smaller numbers. The sequence of primes begins

$$2, \ 3, \ 5, \ 7, \ 11, \ 13, \ 17, \ 19, \ 23, \ 29, \ 31, \ 37, \ldots$$

Note that 1 is excluded, by convention. Prime numbers are of fundamental importance in mathematics because every whole number is a product of primes – for instance,

$$2,007 = 3 \times 3 \times 223$$
$$2,008 = 2 \times 2 \times 2 \times 251$$
$$2,009 = 7 \times 7 \times 41$$

Moreover (only mathematicians worry about this sort of thing, but actually it's kind of important and surprisingly difficult to prove), there is only one way to achieve this, apart from rearranging the order of the prime numbers concerned. For instance, $2,008 = 251 \times 2 \times 2 \times 2$, but that doesn't count as different. This property is called 'unique prime factorisation'.

If you're worried about 1 here, mathematicians consider it to be the product of *no* primes. Sorry, mathematics is like that sometimes.

The primes seem to be scattered fairly unpredictably. Apart from 2, they are all odd – because an even number is divisible by 2, so it can't be prime unless it is equal to 2. Similarly, 3 is the only prime that is a multiple of 3, and so on.

Euclid proved that there is no largest prime. In other words, there exist infinitely many primes. Given any prime p, you can always find a bigger one. In fact, any prime divisor of $p! + 1$ will do the job. Here $p! = p \times (p - 1) \times (p - 2) \times \cdots \times 3 \times 2 \times 1$, a product called the *factorial* of p. For instance,

$$7! = 7 \times 6 \times 5 \times 4 \times 3 \times 2 \times 1 = 5,040.$$

The largest *known* prime is another matter, because Euclid's method isn't a practical way to generate new primes explicitly. As I write, the largest known prime is

$$2^{43,112,609} - 1$$

which has 12,978,189 digits when written out in decimal notation (see page 153).

Twin primes are pairs of primes that differ by 2. Examples are (3, 5), (5, 7), (11, 13), (17, 19), and so on. The twin primes conjecture states that there are infinitely many twin primes. This is widely believed to be true, but has never been proved. Or disproved. The largest known twin primes, to date, are:

$$2,003,663,613 \times 2^{195,000} - 1 \text{ and } 2,003,663,613 \times 2^{195,000} + 1$$

with 58,711 digits each.

Nicely does it ... In 1994 Thomas Nicely was investigating twin primes by computer, and noticed that his results disagreed with previous computations. After spending weeks searching for errors in his program, he traced the problem to a previously unknown bug in the Intel™ Pentium™ microprocessor chip. At

that time, the Pentium was the central processing unit of most of the world's computers. See
www.trnicely.net/pentbug/bugmail1.html

•••

A Little-Known Pythagorean Curiosity

It is well known that any two Pythagorean triples can be combined to yield another one. In fact, if

$$a^2 + b^2 = c^2$$

and

$$A^2 + B^2 = C^2$$

then

$$(aA - bB)^2 + (aB + bA)^2 = (cC)^2$$

However, there is a lesser-known feature of this method for combining Pythagorean triples. If you think of it as a kind of 'multiplication' for triples, then we can define a triple to be *prime* if it is not the product of two smaller triples. Then every Pythagorean triple is a product of distinct prime Pythagorean triples; moreover, this 'prime factorisation' of triples is essentially unique, except for some trivial distinctions which I won't go into here.

It turns out that the prime triples are those for which the hypotenuse is a prime number of the form $4k + 1$ and the other two sides are both non-zero, *or* the hypotenuse is 2 or a prime of the form $4k - 1$ and one of the other sides *is* zero (a 'degenerate' triple).

For instance, the 3–4–5 triple is prime, and so is the 5–12–13 triple, because their hypotenuses are both $4k + 1$ primes. The 0–7–7 triple is also prime. The 33–56–65 triple is not prime – it is the 'product' of the 3–4–5 and 5–12–13 triples.

Just thought you'd like to know.

•••

Digital Century

Place *exactly* three common mathematical symbols between the digits

    1 2 3 4 5 6 7 8 9

so that the result equals 100. The same symbol can be repeated if you wish, but each repeat counts towards your limit of three. Rearranging the digits is not permitted.

Answer on page 266

• •

Squaring the Square

We all know that a rectangular floor can be tiled with square tiles of equal size – provided its edges are integer (whole-number) multiples of the size of the tile. But what happens if we are required to use square tiles which all have *different* sizes?

The first 'squared rectangle' was published in 1925 by Zbigniew Morón, using ten square tiles of sizes 3, 5, 6, 11, 17, 19, 22, 23, 24 and 25.

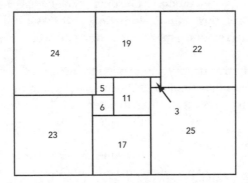

Morón's first squared rectangle.

Not long after, he found a squared rectangle using nine square tiles with sizes 1, 4, 7, 8, 9, 10, 14, 15 and 18. *Can you arrange these tiles to make a rectangle?* As a hint, it has size 32 × 33.

What about making a *square* out of different square tiles? For a long time this was thought to be impossible, but in 1939 Roland Sprague found 55 distinct square tiles that fit together to make a square. In 1940 four mathematicians (Leonard Brooks, Cedric Smith, Arthur Stone and William Tutte, then undergraduates at Trinity College, Cambridge) published a paper relating the problem to electrical networks – the network encodes what size the squares are, and how they fit together. This method led to more solutions.

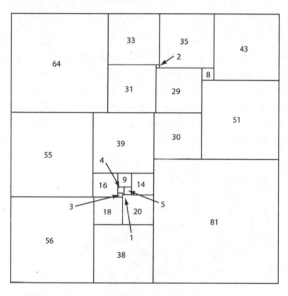

Willcocks's squared square with 24 tiles.

In 1948 Theophilus Willcocks found 24 squares that fit together to make a square. For a while it was thought that no smaller set would do the job, but in 1962 Adrianus Duijvestijn used a computer to show that only 21 square tiles are needed, and this is the minimum number. Their sizes are 2, 4, 6, 7, 8, 9, 11, 15, 16, 17, 18, 19, 24, 25, 27, 29, 33, 35, 37, 42 and 50. *Can you arrange Duijvestijn's 21 tiles to make a square?* As a hint, it has size 112 × 112.

Finally, a really hard one: can you tile the infinite plane, leaving no gaps, using exactly one tile of each whole number size: 1, 2, 3, 4, and so on? This problem remained open until 2008, when Frederick and James Henle proved that you can. See their article 'Squaring the plane', *American Mathematical Monthly*, volume 115 (2008), pages 3–12.

For further information, see **www.squaring.net**

Answers on page 267

• •

Magic Squares

I'm on a bit of a square kick here, so let me mention the most ancient 'square' mathematical recreation of them all. According to a Chinese myth, the Emperor Yu, who lived in the third millennium BC, came across a sacred turtle in a tributary of the Yellow River, with strange markings on its shell. These markings are now known as the *Lo shu* ('Lo river writing').

The *Lo Shu*.

The markings are numbers, and they form a square pattern:

```
4  9  2
3  5  7
8  1  6
```

Here every row, every column and every diagonal adds to the same number, 15. A number square with these properties is said to be *magic*, and the number concerned is its *magic constant*. Usually the square is made from successive whole numbers, 1, 2, 3, 4, and so on, but sometimes this condition is relaxed.

Dürer's *Melancholia* and its magic square.

In 1514 the artist Albrecht Dürer produced an engraving, 'Melancholia', containing a 4 × 4 magic square (top right corner). The middle numbers in the bottom row are 15–14, the date of the work. This square contains the numbers

```
16   3    2   13
 5  10   11    8
 9   6    7   12
 4  15   14    1
```

and has magic constant 34.

Using consecutive whole numbers 1, 2, 3, ..., and counting rotations and reflections of a given square as being the same, there are precisely:

- 1 magic square of size 3 × 3
- 880 magic squares of size 4 × 4
- 27,5305,224 magic squares of size 5 × 5

The number of 6 × 6 magic squares is unknown, but has been estimated by statistical methods to be about 1.77×10^{19}.

The literature on magic squares is gigantic, including many variations such as magic cubes. The website mathworld.wolfram. com/MagicSquare.html is a good place to look, but there are plenty of others.

● ●

Squares of Squares

Magic squares are so well known that I'm not going to say a lot about the common ones, but some of the variants are more interesting. For instance, is it possible to make a magic square whose entries are all distinct perfect squares? Call this a *square of squares*. (Clearly the condition of using *consecutive* whole numbers must be ignored!)

We still have no idea whether a 3×3 square of squares exists. Near misses include Lee Sallows's

$$
\begin{array}{ccc}
127^2 & 46^2 & 58^2 \\
2^2 & 113^2 & 94^2 \\
74^2 & 82^2 & 97^2
\end{array}
$$

for which all rows, columns, and *one* diagonal have the same sum. Another near miss is magic:

$$
\begin{array}{ccc}
373^2 & 289^2 & 565^2 \\
\mathbf{360{,}721} & 425^2 & 23^2 \\
205^2 & 527^2 & \mathbf{222{,}121}
\end{array}
$$

However, only seven entries are square – I've marked the exceptions in bold. It was found by Sallows and (independently) Andrew Bremner.

In 1770 Euler sent the first 4×4 square of squares to Joseph-Louis Lagrange:

$$
\begin{array}{cccc}
68^2 & 29^2 & 41^2 & 37^2 \\
17^2 & 31^2 & 79^2 & 32^2 \\
59^2 & 28^2 & 23^2 & 61^2 \\
11^2 & 77^2 & 8^2 & 49^2
\end{array}
$$

It has magic constant, 8515.

Christian Boyer has found 5×5, 6×6 and 7×7 squares of squares. The 7×7 square uses squares of consecutive integers from 0^2 to 48^2:

25^2	45^2	15^2	14^2	44^2	5^2	20^2
16^2	10^2	22^2	6^2	46^2	26^2	42^2
48^2	9^2	18^2	41^2	27^2	13^2	12^2
34^2	37^2	31^2	33^2	0^2	29^2	4^2
19^2	7^2	35^2	30^2	1^2	36^2	40^2
21^2	32^2	2^2	39^2	23^2	43^2	8^2
17^2	28^2	47^2	3^2	11^2	24^2	38^2

Ring a-Ring a-Ringroad

The M25 motorway completely encircles London, and in Britain we drive on the left. So if you travel clockwise round the M25 you stay on the outside carriageway, whereas travelling anti-clockwise keeps you on the inside carriageway, which is shorter. But how much shorter? The total length of the M25 is 188 km (117 miles), so the advantage of being on the inside carriageway ought to be quite a lot – shouldn't it?

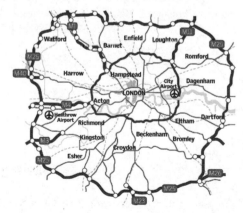

The M25 motorway.

Suppose that two cars travel right round the M25, staying in the outside lane—no, make that two *white vans* travelling right round the M25, staying in the outside lane, as they tend to do. Assume that one is going clockwise and the other anticlockwise, and suppose (which is not entirely true but makes the problem specific) that the distance between these two lanes is always 10 metres. How much further does the clockwise van travel than the anticlockwise one? You may assume that the roads all lie in a flat plane (which also isn't entirely true).

Answer on page 267

Answer on page 267

• •

Pure v. Applied

Relations between pure and applied mathematicians are based on trust and understanding. Pure mathematicians do not trust applied mathematicians, and applied mathematicians do not understand pure mathematicians.

• •

Magic Hexagon

Magic hexagons are like magic squares, but using a hexagon-shaped arrangement of hexagons, like a chunk of a honeycomb:

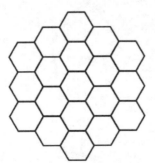

Grid for magic hexagon.

Your task is to place the numbers from 1 to 19 in the hexagons so that any straight line of three, four or five cells, in any of the three directions, add up to the same magic constant, which I can reveal must be 38.

Answer on page 269

• •

Pentalpha

This ancient geometrical puzzle is easy if you look at it the right way, and baffling if you don't.

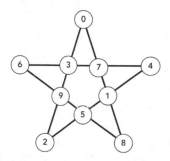

Follow the rules to place nine counters.

You have nine counters, to be placed on the circles in a five-pointed star. Here I've numbered the circles to help explain the solution. In the real game, there aren't any numbers. Successive counters must be positioned by placing them in an empty circle, jumping over an adjacent circle (which may be empty or full) to land on an empty circle adjacent to the one jumped over, so that all three circles involved in the move are on the same straight line. For instance, if circles 7 and 8 are empty, you can place a counter on 7 and jump over 1 to land on 8. Here 1 can be empty or full – it doesn't matter. But you are not allowed to jump 7 over 1 to land on 4 or 5 because now the three circles involved are not in a straight line.

If you try placing counters at random, you usually run out of suitable pairs of empty circles before finishing the puzzle.

Answer on page 270

• •

Wallpaper Patterns

A *wallpaper pattern* repeats the same image in two directions: down the wall and across the wall (or on a slant). The repetition down the wall comes from the paper being printed in a continuous roll, using a revolving cylinder to create the pattern. The repetition across the wall makes it possible to continue the pattern sideways, across adjacent strips of paper, to cover the entire wall. A 'drop' from one panel to the next causes no problems, and can actually make it easier to hang the paper.

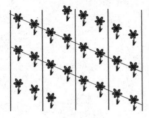

Wallpaper patterns repeat in two directions.

The number of possible *designs* for wallpaper is effectively infinite. But different designs can have the same underlying pattern, it's just that the basic image that gets repeated is different. For instance, the flower in the design above could be replaced by a butterfly, or a bird, or an abstract shape. So mathematicians distinguish *essentially* different patterns by their symmetries. What are the different ways in which we can slide the basic image, or rotate it, or even flip it over (like reflecting it in a mirror), so that the end result is the same as the start?

For my pattern of flowers, the only symmetries are slides along the two directions in which the basic image repeats, or several such slides performed in turn. This is the simplest type of symmetry, but there are more elaborate ones involving rotations and reflections as well. In 1924 George Pólya and Paul Niggli proved that there are exactly 17 different symmetry types of wallpaper pattern – surprisingly few.

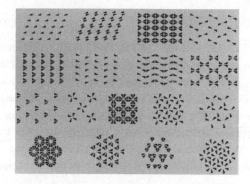

The 17 types of wallpaper pattern.

In three dimensions the corresponding problem is to list all possible symmetry types of atomic lattices of crystals. Here there are 230 types. Curiously, that answer was discovered before anyone solved the much easier two-dimensional version for wallpaper.

. . . .

How Old Was Diophantus?

A few pages back, in the section on Fermat's Last Theorem, I mentioned Diophantus of Alexandria, who lived around AD 250 and wrote a famous book on equations, the *Arithmetica*. That is virtually all we know about him, except that a later source tells us his age – assuming it is authentic. That source says this:

Diophantus's childhood lasted one-sixth of his life. His beard grew after one-twelfth more. He married after one-seventh more. His son was born five years later. The son lived to half his father's age. Diophantus died four years after his son. How old was Diophantus when he died?

Answer on page 271

. . . .

If You Thought Mathematicians were Good at Arithmetic . . .

Ernst Kummer was a German algebraist, who did some of the best work on Fermat's Last Theorem before the modern era. However, he was poor at arithmetic, so he always asked his students to do the calculations for him. On one occasion he needed to work out 9×7. 'Umm . . . nine times seven is . . . nine times . . . seven . . . is . . .'

'Sixty-one,' suggested one student. Kummer wrote this on the blackboard.

'No, Professor! It should be sixty-seven!' said another.

'Come, come, gentlemen,' said Kummer. 'It can't be both. It must be one or the other.'

• •

The Sphinx is a Reptile

Well, a rep-tile, which isn't quite the same thing. Short for 'replicating tile', this word refers to a shape that appears – magnified – when several copies of it are put together. The most obvious rep-tile is a square.

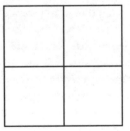

Four square tiles make a bigger square.

However, there are many others, such as these:

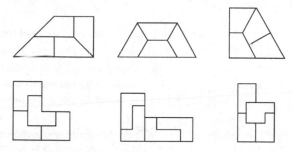

More interesting rep-tiles.

A famous rep-tile is the *sphinx*. Can you assemble four copies of a sphinx to make a bigger sphinx? You can turn some of the tiles over if you want.

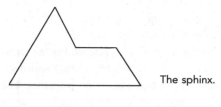

The sphinx.

Answer on page 271

• •

Six Degrees of Separation

In 1998 Duncan Watts and Steven Strogatz published a research paper in the science journal *Nature* about 'small-world networks'. These are networks in which certain individuals are unusually well-connected. The paper triggered a mass of research in which the ideas were applied to real networks such as the Internet and the transmission of disease epidemics.

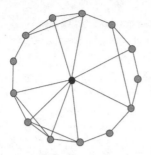

A small-world network. The black individual in the middle is connected to many others, unlike the grey individuals.

The story began in 1967 when the psychologist Stanley Milgram prepared 160 letters with the name of his stockbroker on the envelope – but no address. Then he 'lost' the letters so that random members of the populace could find them, and, he hoped, send them on. Many of the letters duly arrived at the stockbroker's office, and when they did, they had done so in at most six steps. This led Milgram to the idea that we are connected to every other person on the planet by at most five intermediaries – six degrees of separation.

I was explaining the *Nature* paper and its background to my friend Jack Cohen, in the maths common room. Our head of department walked past, stopped, and said, 'Nonsense! Jack, how many steps are there between you and a Mongolian yak herder?' Jack's instant response was: 'One!' He then explained that the person in the office next to his was an ecologist who had worked in Mongolia. This kind of thing happens to Jack, because he is one of those unusually well-connected people who makes small-world networks hang together. For example, he causes both me and my head of department to be only two steps away from a Mongolian yak herder.

You can explore the small-world phenomenon using the Oracle of Bacon, at **oracleofbacon.org**. Kevin Bacon is an actor who has appeared in a lot of movies. Anyone who has appeared in the same movie as Kevin has a *Bacon number* of 1. Anyone who has appeared in the same movie as anyone with a Bacon number of 1 has a Bacon number of 2, and so on. If Milgram is right, virtually every actor (movies being the relevant 'world') has a

Bacon number of 6 or less. At the Oracle, when you type in an actor's name it tells you the Bacon number, and which movies form the links. For instance,

- Michelle Pfeiffer appeared in *Amazon Women on the Moon* in 1987 with:
- David Alan Grier, who appeared in *The Woodsman* in 2004 with:
- Kevin Bacon

so Michelle's Bacon number is 2.

It's not easy to find anyone with Bacon number bigger than 2! One of them is

- Hayley Mooy, who appeared in *Star Wars: Episode III – Revenge of the Sith* in 2005 with:
- Samuel L. Jackson, who appeared in *Snakes on a Plane* in 2006 with:
- Rachel Blanchard, who appeared in *Where the Truth Lies* in 2005 with:
- Kevin Bacon

so Hayley has Bacon number 3.

Mathematicians have their own version of the Oracle of Bacon, centred on the late Paul Erdős. Erdős wrote more joint research papers than any other mathematician, so the game goes the same way but the links are joint papers. My Erdős number is 3, because

- I wrote a joint paper with:
- Marty Golubitsky, who wrote a joint paper with:
- Bruce Rothschild, who wrote a joint paper with:
- Paul Erdős

and no shorter chain exists. One of my former students, who has written a joint paper with me but with no one else, has Erdős number 4.

Usually, more people take part in the same movie than co-author the same mathematics research paper – though I can't say the same about some areas of biology or physics. So you'd expect bigger Erdős numbers than Bacon numbers, on the whole. All mathematicians with Erdős number 1 or 2 are listed at www.oakland.edu/enp

• •

Trisectors Beware!

Euclid tells us how to bisect an angle – divide it into two equal parts – and by repeating this method we can divide any given angle into $4, 8, 16, \ldots, 2^n$ equal parts. But Euclid doesn't explain how to *trisect* an angle – divide it into three equal parts. (Or *quinquisect*, five equal parts, or ....)

Traditionally, Euclidean constructions are carried out using only two instruments – an idealised ruler, with no markings along its edge, to draw a straight line, and an idealised (pair of) compass(es), which can draw a circle. It turns out that these instruments are inadequate for trisecting angles, but the proof had to wait until 1837, when Pierre Wantzel used algebraic methods to show that no ruler-and-compass trisection of the angle 60° is possible. Undeterred, many amateurs continue to seek trisections. So it may be worth explaining why they don't exist.

Any point can be constructed approximately, and the approximation can be as accurate as we wish. To trisect an angle to an accuracy of, say one-trillionth of a degree is easy – in principle. The mathematical problem is not about practical solutions: it is about the existence, or not, of ideal, infinitely accurate ones. It is also about *finite* sequences of applications of ruler and compass: if you allow infinitely many applications, again any point can be constructed – this time exactly.

The key feature of Euclidean constructions is their ability to form square roots. Repeating an operation leads to complicated combinations of square roots of quantities involving square roots of ... well, you get the idea. But that's all you can do with the traditional instruments.

Turning now to algebra, we find the coordinates of such points by starting with rational numbers and repeatedly taking square roots. Any such number satisfies an algebraic equation of a very specific kind. The highest power of the unknown that appears in the equation (called the *degree*) must be the square, or the fourth power, or the eighth power ... That is, the degree must be a power of 2.

An angle of 60° can be formed from three constructible points: (0, 0), (1, 0), and $(\frac{1}{2}, \frac{\sqrt{3}}{2})$, which lie on the unit circle (radius 1, with its centre at the origin of the coordinate system). Trisecting this angle is equivalent to constructing a point (x, y) where the line at 20° to the horizontal axis crosses this circle. Using trigonometry and algebra, the coordinate x of this point is a solution of a *cubic* equation with rational coefficients. In fact x satisfies the equation $8x^3 - 6x - 1 = 0$. But the degree of a cubic is 3, which is not a power of 2. Contradiction – so no trisection is possible. Yes, you can get as close as you wish, but not spot on.

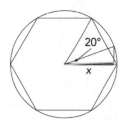

Trisecting 60° is
equivalent to constructing x.

Trisectors often look for the impossible method even though they've heard of Wantzel's proof. They say things like 'I know it's impossible algebraically, but what about *geometrically*?' But Wantzel's proof shows that there is no geometric solution. It uses algebraic *methods* to do that, but algebra and geometry are mutually consistent parts of mathematics.

I always tell would-be trisectors that if they think they've found a trisection, a direct consequence is that 3 is an even number. Do they really want to go down in history as making that claim?

If the conditions of the problem are relaxed, many trisections exist. Archimedes knew of one that used a ruler with just two marks along its edge. The Greeks called this kind of technique a *neusis* construction. It involves sliding the rule so that the marks fall on two given curves – here a line and a circle:

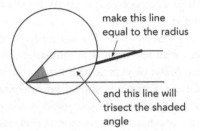

make this line equal to the radius

and this line will trisect the shaded angle

How Archimedes trisected an angle.

Langford's Cubes

The Scottish mathematician C. Dudley Langford was watching his young son playing with six coloured blocks – two of each colour. He noticed that the boy had arranged them so that the two yellow blocks (say) were separated by one block, the two blue blocks were separated by two blocks, and the two red blocks were separated by three blocks. Here I've used white for yellow, grey for blue and black for red to show you what I mean:

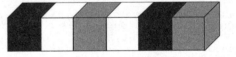

Langford's cubes.

In between the white blocks we find just one block (which happens to be grey). Between the grey blocks are two blocks (one black, one white). And between the black blocks are three blocks (two white and one grey). Langford thought about this and was able to prove that this is the only such arrangement, except for its left–right reversal.

He wondered if you could do the same with more colours – such as four. And he found that again there is only one arrangement, plus its reversal. Can you find it? The simplest way to work on the puzzle is to use playing cards instead of blocks. Take two aces, two 2's, two 3's and two 4's. Can you lay the cards in a row to get exactly one card between the two aces, two cards between the two 2's, three cards between the two 3's and four cards between the two 4's?

There are no such arrangements with five or six pairs of cards, but there are 26 of them with seven pairs. In general, solutions exist if and only if the number of pairs is a multiple of 4 or one less than a multiple of 4. No formula is known for how many solutions there are, but in 2005 Michaël Krajecki, Christophe Jaillet and Alain Bui ran a computer for three months and found that there are precisely 46,845,158,056,515,936 arrangements with 24 pairs.

Answer on page 271

• •

Duplicating the Cube

I'll briefly mention another cube problem: the third famous 'geometric problem of antiquity'. It's nowhere near as well known as the other two – trisecting the angle and squaring the circle. The traditional story is that an altar in the shape of a perfect cube must be doubled in volume. This is equivalent to constructing a line of length $\sqrt[3]{2}$ starting from the rational points of the plane. The desired length satisfies another cubic equation, this time the obvious one, $x^3 - 2 = 0$. For the same reason that trisecting the angle is impossible, so is duplicating the cube, as Pierre Wantzel pointed out in his 1837 paper. Cube-duplicators are so rare that you hardly ever come across one.* Trisectors are ten a penny.

• •

* Though we shouldn't forget Edwin J. Goodwin, whose work on squaring the circle nearly caused a rumpus in Indiana (page 25).

Magic Stars

Here is a five-pointed star. It is a *magic* star because the numbers on any line of four circles add up to the same total, 24. But it's not a very good magic pentacle, because it doesn't use the numbers from 1 to 10. Instead, it uses the numbers from 1 to 12 with 7 and 11 missing.

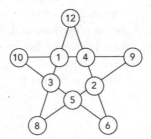

Five-pointed magic star.
Numbers are not consecutive.

It turns out that this is the best you can do with a five-pointed star. But if you use a six-pointed star, it is possible to place the numbers 1 to 12 in the circles, using one of each, so that each line of four numbers has the same total. (As a hint, the total has to be 26.) And, just to make the puzzle harder, I want you to make the six outermost numbers add up to 26 as well.

Where do the numbers go?

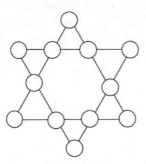

Write the numbers 1 to 12 in the circles to make this star magic.

Answer on page 271

Curves of Constant Width

The circle has the same width in any orientation. If you place it between two parallel lines, it can turn into any position. This is one reason why wheels are circular, and it's why circular logs make excellent rollers.

Is it the *only* curve like that?

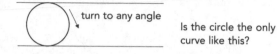

turn to any angle

Is the circle the only curve like this?

Answer on page 272

Connecting Cables

Can you connect the fridge, cooker and dishwasher to the three corresponding electrical sockets, using cables that don't go through the kitchen walls or any of the three appliances, so that no two cables cross?

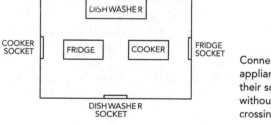

Connect the appliances to their sockets without any crossings.

In ordinary three-dimensional space this puzzle is a bit artificial, but in two dimensions it's a genuine problem, as any inhabitant of Flatland will tell you. A kitchen with no doors is an even bigger problem, but there you go.

Answer on page 272

Coin Swap

The first diagram shows six silver coins, A, C, E, G, I and K, and six gold coins, B, D, F, H, J and L. Your job is to move the coins into the arrangement shown in the second diagram. Each move must swap one silver coin with one adjacent gold coin; two coins are adjacent if there is a straight line joining them. The smallest number of moves that solves this puzzle is known to be 17. Can you find a 17-move solution?

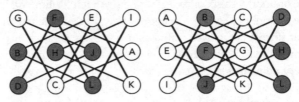

Move the coins from the first position to the second.

Answer on page 273

The Stolen Car

Nigel Fenderbender bought a secondhand car for £900 and advertised it in the local paper for £2,900. A respectable-looking elderly gentleman dressed as a clergyman turned up at the doorstep and enquired about the car, and bought it at the asking price. However, he mistakenly made his cheque out for £3,000, and it was the last cheque in his chequebook.

Now, Fenderbender had no cash in the house, so he nipped next door to the local newsagent, Maggie Zine, who was a friend of his, and got her to change the cheque. He paid the clergyman £100 change. However, when Maggie tried to pay the cheque in at the bank, it bounced. In order to pay back the newsagent, she was forced to borrow £3,000 from another friend, Honest Harry.

After Fenderbender had repaid this debt as well, he complained vociferously: 'I lost £2,000 profit on the car, £100 in change, £3,000 repaying the newsagent and another £3,000 repaying Honest Harry. That's £8,100 altogether!'

How much money had he actually lost?

Answer on page 273

••

Space-Filling Curves

We ordinarily think of a curve as being much 'thinner' than, say, the interior of a square. For a long time, mathematicians thought that since a curve is one-dimensional, and a square is two-dimensional, it must be impossible for a curve to pass through every point inside a square.

Not so. In 1890 the Italian mathematician Giuseppe Peano discovered just such a *space-filling curve*. It was infinitely long and infinitely wiggly, but it still fitted the mathematical concept of a curve – which basically is some kind of bent line. In this case, *very* bent. A year later the German mathematician David Hilbert found another one. These curves are too complicated to draw – and if you could, you might as well just draw a solid black square like the left-hand picture. Mathematicians define space-filling curves using a step-by-step process that introduces more and more wiggles. At each step, the new wiggles are finer than the previous ones. The right-hand picture shows the fifth stage of this process for Hilbert's curve.

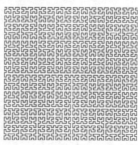

Hilbert's space-filling curve and an approximation.

There is an excellent animation showing successive stages of the construction of the Hilbert curve at en.wikipedia.org/wiki/Hilbert_curve. Similar curves can fill a solid cube, indeed the analogue of a cube in any number of dimensions. So examples like this forced mathematicians to rethink basic concepts, such as 'dimension'. Space-filling curves have been proposed as the basis of an efficient method for searching databases by computer.

• •

Compensating Errors

The class had been given a sum to do, involving three positive whole numbers ('positive' here means 'greater than zero'). During the break, two classmates compared notes.

'Oops. I added the three numbers instead of multiplying them,' said George.

'You're lucky, then,' said Henrietta. 'It's the same answer either way.'

What were the three numbers? What would they have been if there had been only two of them, or four of them, again with their sum equal to their product?

Answer on page 273

• •

The Square Wheel

We seldom see a square wheel, but that's not because such a wheel can't roll without creating a bumpy ride. Circular wheels are great on flat roads. For square wheels, you just need a different shape of road:

A square-wheeled bicycle stays level on this bumpy road.

In fact, the correct shape is a series of inverted arcs of a catenary. A catenary is a U-shaped curve formed by a hanging chain. If the arcs meet at a right angle, a square of the right size will fit snugly, and its centre will stay level as it moves. It turns out that almost any shape of wheel will work provided you build the right kind of road for it to run on. Reinventing the wheel is easy. What counts is reinventing the road.

• •

Why Can't I Divide by Zero?

In general, any number can be divided by any other number – except when the number we are dividing by is 0. 'Division by zero' is forbidden; even our calculators put up error messages if we try it. Why is zero a pariah in the division stakes?

The difficulty is not that we can't *define* division by zero. We could for instance, insist that any number divided by zero gives 42. What we can't do is make that kind of definition and still have all the usual arithmetical rules working properly. With this admittedly very silly definition, we could start from $1/0 = 42$ and apply standard arithmetical rules to deduce that $1 = 42 \times 0 = 0$.

Before we worry about division by zero, we have to agree on the rules that we want division to obey. Division is normally introduced as a kind of opposite to multiplication. What is 6 divided by 2? It is whichever number, multiplied by 2, gives 6. Namely, 3. So the two statements

$$6/2 = 3 \quad \text{and} \quad 6 = 2 \times 3$$

are logically equivalent. And 3 is the only number that works here, so 6/2 is unambiguous.

Unfortunately, this approach runs into major problems when we try to define division by zero. What is 6 divided by 0? It is whichever number, multiplied by 0, gives 6. Namely ... uh. *Any* number multiplied by 0 makes 0; *you can't get 6.*

So 6/0 is out, then. So is any other number divided by 0, except – perhaps – 0 itself. What about 0/0?

Usually, if you divide a number by itself, you get 1. So we could define $0/0$ to be 1. Now $0 = 1 \times 0$, so the relation with multiplication works. Nevertheless, mathematicians insist that $0/0$ doesn't make sense. What worries them is a different arithmetical rule. Suppose that $0/0 = 1$. Then

$$2 = 2 \times 1 = 2 \times (0/0) = (2 \times 0)/0 = 0/0 = 1$$

Oops.

The main problem here is that since any number multiplied by 0 makes 0, we deduce that $0/0$ can be any number whatsoever. If the rules of arithmetic work, and division is the opposite of multiplication, then $0/0$ can take any numerical value. It's not unique. Best avoided, then.

Hang on – if you divide by zero, don't you get *infinity*?

Sometimes mathematicians use that convention, yes. But when they do, they have to check their logic rather carefully, because 'infinity' is a slippery concept. Its meaning depends on the context, and in particular you can't assume that it behaves like an ordinary number.

And even when infinity makes sense, $0/0$ still causes headaches.

• •

River Crossing 2 – Marital Mistrust

Remember Alcuin's letter to Charlemagne (page 20) and the wolf–goat–cabbage puzzle? The same letter contained a more complicated river-crossing puzzle, which may have been invented by the Venerable Bede fifty or so years earlier. It came to prominence in Claude-Gaspar Bachet's seventeenth-century compilation *Pleasant and Delectable Problems*, where it was posed as a problem about jealous husbands who did not trust their wives in the company of other men.

It goes like this. Three jealous husbands with their wives must cross a river, and find a boat with no boatman. The boat can carry only two of them at once. How can they all cross the river so that no wife is left in the company of other men without her husband being present?

Both men and women may row. All husbands are jealous in the extreme: they do not trust their unaccompanied wives to be with another man, *even if the other man's wife is also present.*

Answer on page 273

Answer on page 273

• •

Wherefore Art Thou Borromeo?

Three rings can be linked together in such a way that if any one of them is ignored, the remaining two would pull apart. That is, no two of the rings are linked, only the full set of three. This arrangement is generally known as the *Borromean rings*, after the Borromeo family in Renaissance Italy, who used it as a family emblem. However, the arrangement is much older, and can be found in seventh-century Viking relics. Even in Renaissance Italy, it goes back to the Sforza family; Francesco Sforza permitted the Borromeos to use the rings in their coat of arms as a way of thanking them for their support during the defence of Milan.

Emblem of the Borromeo family and its use (bottom, left of centre) in their coat of arms.

On Isola Bella, one of three islands in Lake Maggiore owned by the Borromeo family, there is a seventeenth-century baroque palazzo built by Vitaliano Borromeo. Here the three-ring emblem can be found in numerous locations, indoors and out. A careful observer (such as a topologist) will discover that the rings depicted there are linked in several topologically distinct ways, only one of which has the key feature that no two rings are linked but all three are:

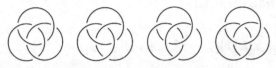

Four variations on the Borromean rings from the family palazzo.

The first version in the picture is the canonical one; it is inset into a floor and also appears in the garden. The second appears on the entrance tickets and on some of the flowerpots. A family crest at the top of the main staircase has the third pattern. Black and white seashells on the floor of a grotto under the palazzo form the fourth pattern. For further information, visit www.liv.ac.uk/~spmr02/rings/

Look at the four versions and explain why they are topologically different.

Can you find an analogous arrangement of four rings, such that if any ring is removed the remaining three can be pulled apart, but the full set of four rings cannot be disentangled?

Answer on page 274

Percentage Play

Alphonse bought two bicycles. He sold one to Bettany for £300, making a loss of 25%, and one to Gemma, also for £300, making a profit of 25%. Overall, did he break even? If not, did he make a profit or a loss, and by how much?

Answer on page 275

Kinds of People

There are 10 kinds of people in the world: those who understand binary numerals, and those who don't.

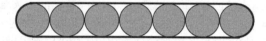

The Sausage Conjecture

This is one of my favourite unsolved mathematical problems, and it is absolutely weird, believe me.

As a warm-up, suppose you are packing a lot of identical circles together in the plane, and 'shrink-wrapping' them by surrounding the lot with the shortest curve you can. With 7 circles, you could try a long 'sausage':

Sausage shape and wrapping.

However, suppose that you want to make the total *area* inside the curve – circles and the spaces between them – as small as possible. If each circle has radius 1, then the area of the sausage is 27.141. But there is a better arrangement of the circles, a hexagon with a central circle, and now the area is 25.533, which is smaller:

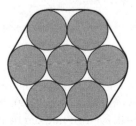

Hexagonal shape and wrapping.

Curiously, if you use identical spheres instead of circles, and shrink-wrap them with the surface of smallest possible area, then for 7 spheres the long sausage shape leads to a smaller total volume than the hexagonal arrangement. This sausage pattern gives the smallest volume inside the wrapping for any number of spheres up to 56. But with 57 spheres or more, the minimal arrangements are more rotund.

Less intuitive still is what happens in spaces of four or more dimensions. The arrangement of 4-dimensional spheres whose wrapping gives the smallest 4-dimensional 'volume' is a sausage for any number of spheres up to 50,000. It's *not* a sausage for 100,000 spheres, though. So the packing of smallest volume uses very long thin strings of spheres until you get an awful lot of them. Nobody knows the precise number at which 4-dimensional sausages cease to be the best.

The really fascinating change *probably* comes at five dimensions. You might imagine that in five dimensions sausages are best for, say, up to 50 billion spheres, but then something more rotund gives a smaller 5-dimensional volume; and for six dimensions the same sort of thing holds up to 29 squillion spheres, and so on. But in 1975 Laszlo Fejes Tóth formulated the *sausage conjecture*, which states that for five or more dimensions, the arrangement of spheres that occupies the smallest volume when shrink-wrapped is *always* a sausage – however large the number of spheres may be.

In 1998 Ulrich Betke, Martin Henk and Jörg Wills proved that Tóth was right for any number of dimensions greater than or equal to 42. To date, that's the best we know.

• •

Tom Fool's Knot

This trick lets you tie a decorative knot while everybody watches. When you challenge them to do the same, they fail. No matter how many times you demonstrate the method, they seem unable to copy it successfully.

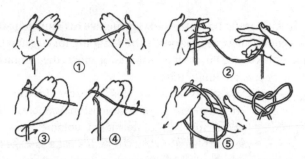

Stages in tying the Tom Fool's knot.

Take a length of soft cord about two metres long and hold it across your palms as in the first diagram, with your hands about half a metre apart. Let the two long ends hang down to counterbalance the weight of the length between the palms. Now bring your hands slowly together, all the while twiddling the fingers of the right hand. The finger twiddles have nothing at all to do with the method of tying the knot, but they distract spectators from the important moves, all of which happen with the left hand. Make the movements of your right hand seem as purposeful as you can.

With the left hand, first slide your thumb under the cord and pick it up, as in the second diagram. Then rapidly withdraw your fingers and replace them behind the hanging end, as shown by the arrow in the second diagram, to reach the position of the third figure. Without stopping, flip your fingers under the horizontal length of cord, as shown by the arrow in the third diagram, and withdraw your thumb. You should now have reached the position shown in the fourth diagram. Finally, use the tips of the fingers of each hand to grasp the end of cord hanging from the other hand, as in the fifth diagram. Holding on to the cord, pull the hands apart, and the lovely symmetrical knot of the final diagram appears.

Practice the method until you can perform it as a single, rhythmic movement. The knot unties if you just pull on the ends

of the cord, so you can tie it over and over again. The trick
becomes more mysterious every time you do it.

•••

New Merology

*Let him that hath understanding count the number of the beast; for it
is the number of a man; and his number is Six hundred threescore and
six.* Revelation of St John 13:18

Or maybe not. The Oxyrhynchus Papyri – ancient documents
found at Oxyrhynchus in Upper Egypt – include a fragment of
the Book of Revelation from the third or fourth century which
contains the earliest known version of some sections. The
number that this papyrus assigns to the Beast is 616, not 666. So
much for barcodes being symbols of evil.* No matter, for this
puzzle is not about the Beast. It is about an idea that its inventor,
Lee Sallows, calls 'new merology'. Let me make it clear that his
proposal is not serious, except as a mathematical problem.†

The traditional method for assigning numbers to names,
known as *gematria*, sets $A = 1$, $B = 2$ up to $Z = 26$. Then you add
up all the numbers corresponding to the letters in the name. But
there are lots of different systems of this kind, and lots of
alphabets. Sallows suggested a more rational method based on
words that denote numbers. For instance, with the numbering
just described, the word ONE becomes $15 + 14 + 5 = 34$. However,

---

* The middle of a supermarket barcode bears lines that would
represent the number 666, except that they have an entirely
different function – they are 'guard bars' that help to correct
errors. Each guard bar has the binary pattern 101, which on a
barcode represents 6. Whence 666. Except that genuine barcode
numbers actually have *seven* binary digits, so that 6 is 1010000,
and ... oh well. This led some American fundamentalists to
denounce barcodes as the work of the Devil. Since it now seems
that the number of the Beast is actually 616, even the numerology
is dodgy.

† I really shouldn't need to say this–but given the previous footnote
...

the number corresponding to ONE surely ought to be 1. Worse, *no* English number word denotes its numerological total, a property we will call 'perfect'.

Sallows wondered what happens if you assign a whole number to each letter, so that as many as possible of the number-words ONE, TWO, and so on are perfect. To make the problem interesting, different letters must be given different values. So you get a whole pile of equations like

$$O + N + E = 1$$
$$T + W + O = 2$$
$$T + H + R + E + E = 3$$

in algebraic unknowns O, N, E, T, W, H, R, ... And you must solve them in integers, all distinct.

The equation $O + N + E = 1$ tells us that some of the numbers have to be negative. Suppose, for example, that $E = 1$ and $N = 2$. Then the equation for ONE tells us that $O = -2$, and similar equations with other number-words imply that $I = 4$, $T = 7$ and $W = -3$. To make THREE perfect we must assign values to H and R. If $H = 3$, then R has to be -9. FOUR involves two more new letters, F and U. If $F = 5$, then $U = 10$. Now, $F + I + V + E = 5$ leads to $V = -5$. Since SIX contains two new letters, we try SEVEN first, which tells us that $S = 8$. Then we can fill in X from SIX, getting $X = -6$. The equation for EIGHT leads to $G = -7$. Now all the number names from ONE to TEN are perfect.

The only extra letter in ELEVEN and TWELVE is L. Remarkably, $L = 11$ makes them *both* perfect. But $T + H + I + R + T + E + E + N = 7 + 3 + 4 + (-9) + 7 + 1 + 1 + 2$, which is 16, so we get stuck at this point.

In fact, we always get stuck at this point: if THIRTEEN is perfect, then

THREE + TEN = THIRTEEN

and we can remove common letters from both sides. This leads to $E = I$, violating the rule that different letters get different values.

However, we can go the other way and try to make ZERO perfect as well as ONE to TWELVE. Using the choices above, $Z + E + R + O = 0$ leads to $Z = 10$, but that's the same as U.

Can you find a different assignment of positive or negative whole-number values to letters, so that all words from ZERO to TWELVE are perfect?

Answer on page 275

Numerical Spell

Lee Sallows also applied new merology to magic, inventing the following trick. Select any number on the board shown below. Spell it out, letter by letter. Add together the corresponding numbers (subtracting those on black squares, adding those on white squares). The result will always be plus or minus the number you chose. For instance, TWENTY-TWO leads to

$$20 - 25 - 4 - 2 + 20 + 11 + 20 - 25 + 7 = 22$$

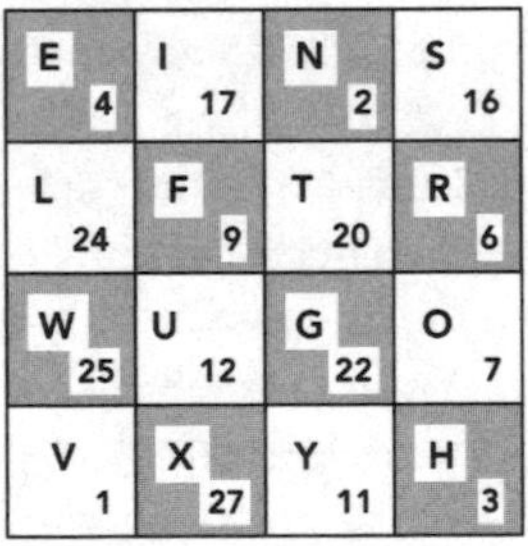

E 4	I 17	N 2	S 16
L 24	F 9	T 20	R 6
W 25	U 12	G 22	O 7
V 1	X 27	Y 11	H 3

Board for Lee Sallows's magic trick.

Spelling Mistakes

'Thare are five mistukes im this centence.'
   True or false?

   Answer on page 275

● ●

Expanding Universe

The starship *Indefensible* starts from the centre of a spherical universe of radius 1,000 light years, and travels radially at a speed of one light year per year – the speed of light. How long will it take to reach the edge of the universe? Clearly, 1,000 years. Except that I forgot to tell you that this universe is expanding. Every year, the universe expands its radius *instantly* by precisely 1,000 light years. Now, how long will it take to reach the edge? (Assume that the first such expansion happens exactly one year after the *Indefensible* starts its voyage, and successive expansions occur at intervals of exactly one year.)

It might seem that the *Indefensible* never gets to the edge, because that is receding faster than the ship can move. But at the instant that the universe expands, the ship is carried along with the space in which it sits, so its distance from the centre expands proportionately. To make these conditions clear, let's look at what happens for the first few years.

In the first year the ship travels 1 light year, and there are 999 light years left to traverse. Then the universe instantly expands to a radius of 2,000 light years, and the ship moves with it. So it is then 2 light years from the centre, and has 1,998 left to travel.

In the next year it travels a further light year, to a distance of 3 light years, leaving 1,997. But then the universe expands to a radius of 3,000 light-years, multiplying its radius by 1.5, so the ship ends up 4.5 light years from the centre, and the remaining distance increases to 2,995.5 light years.

Does the ship ever get to the edge? If so, how long does it take?

Hint: it will be useful to know that the nth harmonic number

$$H_n = 1 + \frac{1}{2} + \frac{1}{3} + \frac{1}{4} + \cdots + \frac{1}{n}$$

is approximately equal to

$$\log n + \gamma$$

where γ is Euler's constant, which is roughly $0.577\,215\,664\,9$.

Answer on page 276

• •

What is the Golden Number?

The ancient Greek geometers discovered a useful idea which they called 'division in extreme and mean ratio'. By this they meant a line AB being cut at a point P, so that the ratios AP : AB and PB : AP are the same. Euclid used this construction in his work on regular pentagons, and I'll shortly explain why. But first, since nowadays we have the luxury of replacing ratios by numbers, let's turn the geometric recipe into algebra. Take PB to be of length 1, and let AP = x, so that AB = $1 + x$. Then the required condition is

$$\frac{1 + x}{x} = \frac{x}{1}$$

so that $x^2 - x - 1 = 0$. The solutions of this quadratic equation are

$$\phi = \frac{1 + \sqrt{5}}{2} = 1.618\,034\ldots$$

and

$$1 - \phi = \frac{1 - \sqrt{5}}{2} = -0.618\,034\ldots$$

Here the symbol ϕ is the Greek letter phi. The number ϕ, known as the *golden number*, has the pleasant property that its reciprocal is

$$\frac{1}{\phi} = \frac{-1 + \sqrt{5}}{2} = 0.618\,034\ldots = \phi - 1$$

The golden number, in its geometric form as 'division in extreme and mean ratio', was the starting point for the Greek geometry of regular pentagons and anything associated with these, such as the dodecahedron and the icosahedron. The connection is this: if you draw a pentagon with sides equal to 1, then the long diagonals have length ϕ:

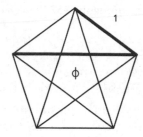

How ϕ appears in a regular pentagon.

The golden ratio is often associated with aesthetics; in particular, the 'most beautiful' rectangle is said to be one whose sides are in the ratio $\phi:1$. The actual evidence for such statements is weak. Moreover, various methods of presenting numerical data exaggerate the role of the golden ratio, so that it is possible to 'deduce' the presence of the golden ratio in data that bear no relation to it. Similarly, claims that famous ancient buildings such as the Great Pyramid of Khufu or the Parthenon were designed using the golden ratio are probably unfounded. As with all numerology, you can find whatever you are looking for if you try hard enough. (Thus 'Parthenon' has 8 letters, 'Khufu' has 5, and $8/5 = 1.6$ – very close to ϕ.[*])

Another common fallacy is to suppose that the golden ratio occurs in the spiral shell of a nautilus. This beautiful shell is – to great accuracy – a type of spiral called a logarithmic spiral. Here each successive turn bears a fixed ratio to the previous one. There is a spiral of this kind for which this ratio equals the golden ratio. But the ratio observed in the nautilus is *not* the golden ratio.

[*] Well, actually 'Parthenon' has 9 letters, but for a moment I had you there. And 1.8 is a lot closer to ϕ than many alleged instances of this number.

The nautilus shell is a logarithmic spiral but its growth rate is not the golden number.

The term 'golden number' is relatively modern. According to the historian Roger Herz-Fischler, it was first used by Martin Ohm in his book 1835 *Die Reine Elementar-Mathematik* ('Pure Elementary Mathematics') as *Goldene Schnitt* ('golden section'). It does not go back to the ancient Greeks.

The golden number is closely connected with the famous Fibonacci numbers, which come next.

● ●

What are the Fibonacci Numbers?

Many people met the Fibonacci numbers for the first time in Dan Brown's bestseller *The Da Vinci Code*. These numbers have a long and glorious mathematical history, which has very little overlap with anything mentioned in the book.

It all began in 1202 when Leonardo of Pisa published the *Liber Abbaci*, or 'Book of Calculation', an arithmetic text which concentrated mainly on financial computations and promoted the use of Hindu-Arabic numerals – the forerunner of today's familiar system, which uses just ten digits, 0 to 9, to represent all possible numbers.

One of the exercises in the book seems to have been Leonardo's own invention. It goes like this: 'A man put a pair of rabbits in a place surrounded on all sides by a wall. How many pairs of rabbits are produced from that pair in a year, if it is supposed that every month each pair produces a new pair, which from the second month onwards becomes productive?'

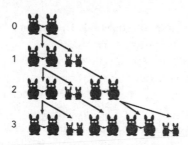

Family tree of of Fibonacci's rabbits.

Say that a pair is *mature* if it can produce offspring, and *immature* if not.

At the start, month 0, we have 1 mature pair.

At month 1 this pair produces an immature pair, so we have 1 mature pair and 1 immature pair – 2 altogether.

At month 2 the mature pair produces another immature pair; the immature pair matures but produces nothing. So now we have 2 mature pairs and 1 immature pair – 3 in total.

At month 3 the 2 mature pairs produce 2 more immature pairs; the immature pair matures but produces nothing. So now we have 3 mature pairs and 2 immature pairs – 5 in total.

At month 4 the 3 mature pairs produce 3 more immature pairs; the 2 immature pairs mature but produce nothing. So now we have 5 mature pairs and 3 immature pairs – 8 in total.

Continuing step by step, we obtain the sequence

$$1, \ 2, \ 3, \ 5, \ 8, \ 13, \ 21, \ 34, \ 55, \ 89, \ 144, \ 233, \ 377$$

for months 0, 1, 2, 3, ..., 12. Here each term after the second is the sum of the previous two. So the answer to Leonardo's question is 377.

Some time later, probably in the eighteenth century, Leonardo was given the nickname Fibonacci – 'son of Bonaccio'. This name was more catchy than Leonardo Pisano Bigollo, which is what he used, so nowadays he is generally known as Leonardo Fibonacci, and his sequence of numbers is known as the *Fibonacci sequence*. The usual modern convention is to put the numbers 0, 1 in front, giving

$$0, \ 1, \ 1, \ 2, \ 3, \ 5, \ 8, \ 13, \ 21, \ 34, \ 55, \ 89, \ 144, \ 233, \ 377$$

although sometimes the initial 0 is omitted. The symbol for the nth Fibonacci number is F_n, starting from $F_0 = 0$.

The Fibonacci numbers as such are pretty useless as a model of the growth of real rabbit populations, although more general processes of a similar kind, called Leslie models, are used to understand the dynamics of animal and human populations. Nevertheless, the Fibonacci numbers are important in several areas of mathematics, and they also turn up in the natural world – though less widely than is often suggested. Extensive claims have been made for their occurrence in the arts, especially architecture and painting, but here the evidence is mostly inconclusive, except when Fibonacci numbers are used deliberately – for instance, in the architect Le Corbusier's 'modulor' system.

The Fibonacci numbers have strong connections with the golden number, which you'll recall is

$$\phi = \frac{1 + \sqrt{5}}{2} = 1.618034 \ldots$$

Ratios of successive Fibonacci numbers, such as 8/5, 13/8, 21/13, and so on become ever closer to ϕ as the numbers get bigger. Or, as mathematicians would say, $> F_{n+1}/F_n$ tends to ϕ as n tends to infinity. For instance, $377/233 = 1.618025 \ldots$. In fact, for integers of a given size, these Fibonacci fractions provide the best possible approximations to the golden number. There is even a formula for the nth Fibonacci number in terms of ϕ:

$$F_n = \frac{\phi^n - (1 - \phi)^n}{\sqrt{5}}$$

and this implies that F_n is the integer closest to $\phi^n/\sqrt{5}$.

If you make squares whose sides are the Fibonacci numbers they fit together very tidily, and you can draw quarter-circles in them to create an elegant *Fibonacci spiral*. Because F_n is close to ϕ^n, this spiral is very close to a logarithmic spiral, which grows in size by ϕ every quarter-turn. Contrary to many claims, this spiral is *not* the same shape as the Nautilus shell's spiral. Look at the picture on page 98 – the Nautilus is more tightly wound.

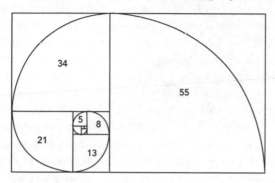

Fibonacci spiral.

However, there is a genuine – and striking – occurrence of Fibonacci numbers in living creatures, namely plants. The flowers of surprisingly many species have a Fibonacci number of petals. Lilies have 3 petals, buttercups have 5, delphiniums have 8, marigolds have 13, asters have 21, and most daisies have 34, 55 or 89. Sunflowers often have 55, 89 or 144.

Other numbers of petals do occur, but much less frequently. Mostly these are twice a Fibonacci number, or a power of 2. Sometimes numbers are from the related *Lucas sequence*:

1, 3, 4, 7, 11, 18, 29, 47, 76, 123, . . .

where again each number after the second is the sum of the previous two, but the start of the sequence is different.

There seem to be genuine biological reasons for these numbers to occur. The strongest evidence can be seen in the heads of daisies and sunflowers, when the seeds have formed. Here the seeds arrange themselves in spiral patterns:

The head of
a daisy.

In the daisy illustrated, the eye sees one family of spirals that
twist clockwise, and a second family of spirals that twist
anticlockwise. There are 21 clockwise spirals and 34 anticlockwise spirals – successive Fibonacci numbers. Similar
numerical patterns, also involving successive Fibonacci numbers,
occur in pine cones and pineapples.

The precise reasons for Fibonacci numerology in plant life are
still open to debate, though a great deal is understood. As the tip
of the plant shoot grows, long before the flowers appear, regions
of the shoot form tiny bumps, called primordia, from which the
seeds and other key parts of the flower eventually grow.
Successive primordia form at angles of 137.5° – or 222.5° if we
subtract this from 360°, measuring it the other way round. This is
a fraction $\phi - 1$ of the full circle of 360°. This occurrence of the
golden ratio can be predicted mathematically if we assume that
the primordia are packed as efficiently as possible. In turn,
efficient packing is a consequence of elastic properties of the
growing shoot – the forces that affect the primordia. The genetics
of the plant is also involved. Of course, many real plants do not
quite follow the ideal mathematical pattern. Nevertheless, the
mathematics and geometry associated with the Fibonacci
sequence provide significant insights into these numerical
features of plants.

The Plastic Number

The plastic number is a little-known relative of the famous golden number. We've just seen how the Fibonacci numbers create a spiralling system of squares, related to the golden number. There is a similar spiral diagram for the plastic number, but composed of equilateral triangles. In the diagram below, the initial triangle is marked in black and successive triangles spiral in a clockwise direction: the spiral shown is again roughly logarithmic. In order to make the shapes fit, the first three triangles all have side 1. The next two have side 2, and then the numbers go 4, 5, 7, 9, 12, 16, 21, and so on.

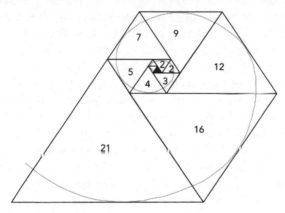

Padovan spiral.

Again there is a simple rule for finding these numbers, analogous to that for Fibonacci numbers: each number in the sequence is the sum of the previous number *but one*, together with the one before that. For example,

$$12 = 7 + 5, 16 = 9 + 7, 21 = 12 + 9$$

This pattern follows from the way the triangles fit together. If P_n is the nth Padovan number (starting from $P_0 = P_1 = P_2 = 1$), then

$$P_n = P_{n-2} + P_{n-3}$$

The first twenty numbers in the sequence are:

$$1, \; 1, \; 1, \; 2, \; 2, \; 3, \; 4, \; 5, \; 7, \; 9, \; 12, \; 16,$$
$$21, \; 28, \; 37, \; 49, \; 65, \; 86, \; 114, \; 151$$

I call this sequence the *Padovan numbers* because the architect Richard Padovan told me about them, although he denies any responsibility. Curiously, 'Pádova' is the Italian form of 'Padua', and Fibonacci was from Pisa, roughly a hundred miles away. I am tempted to rename the Fibonacci numbers 'Pisan numbers' to reflect the Italian geography, but as you can see I managed to resist.

The *plastic number*, which I denote by p, is roughly 1.324718. It is related to the Padovan numbers in the same way that the golden number is related to the Fibonacci numbers. That is, ratios of successive Padovan numbers, such as 49/37 or 151/114, give good approximations to the plastic number. The pattern of the sequence of Padovan numbers leads to the equation $p^3 - p - 1 = 0$, and p is the unique real solution of this cubic equation. The Padovan sequence increases much more slowly than the Fibonacci sequence because p is smaller than ϕ. There are many interesting patterns in the Padovan sequence. For example, the diagram shows that $21 = 16 + 5$, because triangles adjacent along a suitable edge have to fit together; similarly, $16 = 12 + 4, 12 = 9 + 3$, and so on. Therefore

$$P_n = P_{n-1} + P_{n-5}$$

which is an alternative rule for deriving further terms of the sequence. This equation implies that $p^5 - p^4 - 1 = 0$, and it is not immediately obvious that p, defined as a solution of a cubic equation, must also satisfy this *quintic* (fifth-degree) equation.

• •

Family Occasion

'It was a wonderful party,' said Lucilla to her friend Harriet.

'Who was there?'

'Well – there was one grandfather, one grandmother, two fathers, two mothers, four children, three grandchildren, one brother, two sisters, two sons, two daughters, one father-in-law, one mother-in-law and one daughter-in-law.'

'Wow! Twenty-three people!'

'No, it was less than that. A lot less.'

What is the *smallest* size of party that is consistent with Lucilla's description?

Answer on page 277

Don't Let Go!

Topology is a branch of mathematics in which two shapes are 'the same' if one can be continuously deformed into the other. So you can bend, stretch, and shrink, but not cut. This ancient topological chestnut still has many attractions – in particular, not everyone has seen it before. What you have to do is pick up a length of rope, with the left hand holding one end and the right hand holding the other, and tie a knot in the rope *without letting go of the ends*.

Answer on page 277

Theorem: All Numbers are Interesting

Proof: For a contradiction, suppose not. Then there is a smallest uninteresting number. But being the *smallest* one singles it out among all other numbers, making it special, hence interesting – contradiction.

Theorem: All Numbers are Boring

Proof: For a contradiction, suppose not. Then there is a smallest non-boring number.

And your point is—?

● ●

The Most Likely Digit

If you look at a list of numerical data, and count how often a given digit turns up as the *first* digit in each entry, which digit is most likely? The obvious guess is that every digit has the same chance of occurring as any other. But it turns out that for most kinds of data, this is wrong.

Here's a typical data set – the areas of 18 islands in the Bahamas. I've given the figures in square miles and in square kilometres, for reasons I'll shortly explain.

Island	Area (sq mi)	Area (sq km)
Abaco	649	1681
Acklins	192	497
Berry Islands	2300	5957
Bimini Islands	9	23
Cat Island	150	388
Crooked and Long Cay	93	241
Eleuthera	187	484
Exuma	112	290
Grand Bahama	530	1373
Harbour Island	3	8
Inagua	599	1551
Long island	230	596
Mayaguana	110	285
New Providence	80	207
Ragged Island	14	36
Rum Cay	30	78
San Salvador	63	163
Spanish Wells	10	26

For the square mile data, the number of times that a given first digit (shown in brackets) occurs goes like this:

(1) 7 (2) 2 (3) 2 (4) 0 (5) 2 (6) 2 (7) 0 (8) 1 (9) 2

and 1 wins hands down. In square kilometres, the corresponding numbers are

(1) 4 (2) 6 (3) 2 (4) 2 (5) 2 (6) 0 (7) 1 (8) 1 (9) 0

and now 2 wins, but only just.

In 1938 the physicist Frank Benford observed that for long enough lists of data, the numbers encountered by physicists and engineers are most likely to start with the digit 1 and least likely to start with 9. The frequency with which a given initial digit occurs – that is, the probability that the first digit takes a given value – *decreases* as the digits increase from 1 to 9. Benford discovered empirically that the probability of encountering n as the first decimal digit is

$$\log_{10}(n+1) - \log_{10}(n)$$

where the subscript 10 means that the logarithms are to base ten. (The value $n = 0$ is excluded because the initial digit is by definition the first non-zero digit.) Benford called this formula the law of anomalous numbers, but nowadays it's usually known as *Benford's law*.

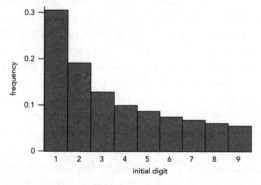

Theoretical frequencies according to Benford's law.

For the Bahama Island data, the frequencies look like this:

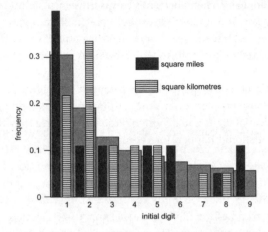

Observed frequencies for the areas of the Bahama Islands, compared with Benford's theoretical ideal.

There are some differences between theory and reality, but the data sets here are fairly small, so we would expect that. Even with only 18 numbers, there is a strong prevalence of 1's and 2's – which, according to Benford's law, should between them occur just over half the time.

Benford's formula is far from obvious, but a little thought shows that the nine frequencies are unlikely to be identical. Think of a street of houses, numbered from 1 upwards. The probability of a given digit coming first varies considerably with the number of houses on the street. If there are nine houses, each digit occurs equally often. But, if there are 19, then the initial digit is 1 for houses 1 and 10–19, a frequency of 11/19, or more than 50%. As the length of the street increases, the frequency with which a given first digit occurs wanders up and down in a complicated but computable manner. The nine frequencies are the same *only* when the number of houses is 9, 99, 999, and so on.

Benford's formula is distinguished by a beautiful property: it is *scale-invariant*. If you measure the areas of Bahamian islands in square miles or square kilometres, if you multiply house numbers by 7 or 93, then – provided you have a big enough sample – the same law applies. In fact, Benford's Law is the *only* scale-invariant frequency law. It is unclear why nature prefers scale-invariant frequencies, but it seems reasonable that the natural world should not be affected by the units in which humans choose to measure it.

Tax collectors use Benford's law to detect fake figures in tax forms, because people who invent fictitious numbers tend to use the same initial digits equally often. Probably because they think that's what should happen for genuine figures!

• •

Why Call It a Witch?

Maria Agnesi was born in 1718 and died in 1799. She was the daughter of a wealthy silk merchant, Pietro Agnesi (often wrongly said to have been a professor of mathematics at Bologna), and the eldest of his 21 children. Maria was precocious, and published an essay advocating higher education for women when she was nine years old. The essay was actually written by one of her tutors, but she translated it into Latin and delivered it from memory to an academic gathering in the garden of the family home. Her father also arranged for her to debate philosophy in the presence of prominent scholars and public figures. She disliked making a public spectacle of herself and asked her father for permission to become a nun. When he refused, she extracted an agreement that she could attend church whenever she wished, wear simple clothing, and be spared from all public events and entertainments.

Maria Gaetana Agnesi.

From that time on, she focused on religion and mathematics. She wrote a book on differential calculus, printed privately around 1740. In 1748 she published her most famous work, *Instituzioni Analitiche ad Uso Della Gioventù Italiana* ('Analytical Institutions for the Use of the Youth of Italy'). In 1750 Pope Benedict XIV invited her to become professor of mathematics at the University of Bologna, and she was officially confirmed in the role, but she never actually attended the university because this would not have been in keeping with her humble lifestyle. As a result, some sources say she was a professor and others say she wasn't. Was she, or wasn't she? Yes.

There is a famous curve, called the 'witch of Agnesi', which has the equation

$$y = a^3/(a^2 + x^2)$$

where a is a constant. The curve looks remarkably *unlike* a witch – it isn't even pointy:

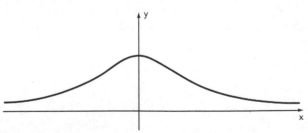

Witch of Agnesi.

So how did this strange name get attached to the curve?
Fermat was the first to discuss this curve, in about 1700.
Maria Agnesi wrote about the curve in her book *Instituzione Analitiche*. The word 'witch' was a mistake in translation. In 1718 Guido Grandi named the curve 'versoria', a Latin term for a rope that turns a sail, because that's what it looked like. In Italian this term became 'versiera', which is what Agnesi called it. But John Colson, who translated various mathematics books into English, mistook 'la versiera' for 'l'aversiera', meaning 'the witch'.

It could have been worse. Another meaning is 'she-devil'.

• •

Möbius and His Band

There are some pieces of mathematical folklore that you really should be reminded about, even though they're 'well known' – just in case. An excellent example is the Möbius band.

Augustus Möbius was a German mathematician, born 1790, died 1868. He worked in several areas of mathematics, including geometry, complex analysis and number theory. He is famous for his curious surface, the *Möbius band*. You can make a Möbius band by taking a strip of paper, say 2 cm wide and 20 cm long, bending it round until the ends meet, then twisting one end through 180°, and finally gluing the ends together. For comparison, make a cylinder in the same way, omitting the twist.

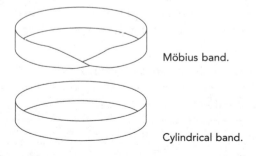

Möbius band.

Cylindrical band.

The Möbius band is famous for one surprising feature: it has only one side. If an ant crawls around on a cylindrical band, it can cover only half the surface – one side of the band. But if an ant crawls around on the Möbius band, it can cover the entire surface. The Möbius band has only one side.

You can check these statements by painting the band. You can paint the cylinder so that one side is red and the other is blue, and the two sides are completely distinct, even though they are separated only by the thickness of the paper. But if you start to paint the Möbius band red, and keep going until you run out of band to paint, the *whole thing* ends up red.

In retrospect, this is not such a surprise, because the 180° twist connects each side of the original paper strip to the other. If you don't twist before gluing, the two sides stay separate. But until Möbius (and a few others) thought this one up, mathematicians didn't appreciate that there are two distinct kinds of surface: those with two sides, and those with one. This turned out to be important in topology. And it showed how careful you have to be about making 'obvious' assumptions.

There are lots of Möbius band recreations. Here are three.

- If you cut the cylindrical band along the middle with scissors, it falls apart into two cylindrical bands. What happens if you try this with a Möbius band?
- Repeat, but this time make the cut about one-third of the way across the width of the band. Now what happens to the cylinder, and to the band?
- Make a band like a Möbius band but with a 360° twist. How many sides does it have? What happens if you cut it along the middle?

The Möbius band is also known as a Möbius strip, but this can lead to misunderstandings, as in a Limerick written by science fiction author Cyril Kornbluth:

A burleycue dancer, a pip
Named Virginia, could peel in a zip;
    But she read science fiction
    and died of constriction
Attempting a Möbius strip.

A more politically correct Möbius limerick, which gives away one of the answers, is:

A mathematician confided
That a Möbius strip is one-sided.
    You'll get quite a laugh
    if you cut it in half,
For it stays in one piece when divided.

Answers on page 277

• •

Golden Oldie

Why did the chicken cross the Möbius band?
    To get to the other ... um ...

• •

Three More Quickies

(1) If five dogs dig five holes in five days, how long does it take ten dogs to dig ten holes? Assume that they all dig at the same rate all the time and all holes are the same size.

(2) A woman bought a parrot in a pet-shop. The shop assistant, who always told the truth, said, 'I guarantee that this parrot will repeat every word it hears.' A week later, the woman took the parrot back, complaining that it hadn't spoken a single word. 'Did anyone talk to it?' asked the suspicious assistant. 'Oh, yes.' What is the explanation?

(3) The planet Nff-Pff in the Anathema Galaxy is inhabited by precisely two sentient beings, Nff and Pff. Nff lives on a large

continent, in the middle of which is an enormous lake. Pff lives on an island in the middle of the lake. Neither Nff nor Pff can swim, fly or teleport: their only form of transport is to walk on dry land. Yet each morning, one walks to the other's house for breakfast. Explain.

Answers on page 277

. . . .

Miles of Tiles

Bathroom walls and kitchen floors provide everyday examples of tiling patterns, using real tiles, plastic or ceramic. The simplest pattern is made from identical square tiles, fitted together like the squares of a chessboard. Over the centuries, mathematicians and artists have discovered many beautiful tilings, and mathematicians have gone a stage further by seeking all possible tilings with particular features.

For instance, exactly three regular polygons tile the entire infinite plane – that is, identical tiles of that shape cover the plane without overlaps or gaps. These polygons are the equilateral triangle, the square and the hexagon:

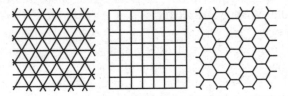

The three regular polygons that tile the plane.

We can be confident that no other *regular* polygon tiles the plane, by thinking about the angles at which the edges of the tiles meet. If several tiles meet at a given point, the angles involved must add to 360°. So the angle at the corner of a tile is 360° divided by a whole number, say $360/m$. As m gets larger, this angle gets smaller. In contrast, as the number of sides of a regular polygon increases, the angle at each corner gets bigger. The effect of this is to 'sandwich' m within very narrow limits, and this in turn restricts the possible polygons.

The details go like this. When $m = 1, 2, 3, 4, 5, 6, 7$, and so on, $360/m$ takes the values 360, 180, 120, 90, 72, 60, $51\frac{3}{7}$, and so on. The angle at the corner of a regular n-gon, for $n = 3, 4, 5, 6, 7$, and so on, is 60, 90, 108, 120, $128\frac{4}{7}$, and so on. The only places where these lists coincide are when $m = 3, 4$, and 6; here $n = 6, 4$ and 3.

Actually, this proof as stated has a subtle flaw. *What have I forgotten to say?*

The most striking omission from my list is the regular pentagon, which does not tile the plane. If you try to fit regular pentagonal tiles together, they don't fit. When three of them meet at a common point, the total angle is $3 \times 108° = 324°$, less than $360°$. But if you try to make four of them meet, the total angle is $4 \times 108° = 432°$, which is too big.

Irregular pentagons can tile the plane, and so can innumerable other shapes. In fact, 14 distinct types of convex pentagon are known to tile the plane. It is probable, but not yet proved, that there are no others. You can find all 14 patterns at www.mathpuzzle.com/tilepent.html
mathworld.wolfram.com/PentagonTiling.html

The mathematics of tilings has important applications in crystallography, where it governs how the atoms in a crystal can be arranged, and what symmetries can occur. In particular, crystallographers know that the possible rotational symmetries of a regular lattice of atoms is tightly constrained. There are 2-fold, 3-fold, 4-fold and 6-fold symmetries – meaning that the arrangement of atoms looks identical if the whole thing is rotated through $\frac{1}{2}$, $\frac{1}{3}$, $\frac{1}{4}$ or $\frac{1}{6}$ of a full turn ($360°$). However, 5-fold symmetries are impossible – just as the regular pentagon cannot tile the plane.

There the matter stood until 1972, when Roger Penrose discovered a new type of tiling, using two types of tile, which he called *kites* and *darts*:

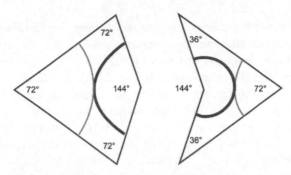

A kite (left) and dart (right). The matching rules require the thick and thin arcs to meet at any join – see the pictures below.

These shapes are derived from the regular pentagon, and the associated tilings are required to obey certain 'matching rules' where several tiles meet, to avoid simple repetitive patterns. Under these conditions, the two shapes can tile the plane, but not by forming a repetitive lattice pattern. Instead, they form a bewildering variety of complicated patterns. Precisely two of these, called the star and sun patterns, have exact fivefold rotational symmetry.

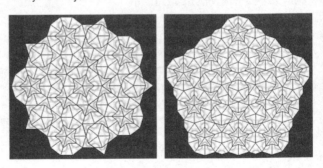

The two fivefold-symmetric Penrose tilings. Left: star pattern; right: sun pattern. Tinted lines illustrate the matching rules. Black lines are edges of tiles.

It then turned out that nature knows this trick. Some chemical compounds can form 'quasicrystals' using Penrose patterns for their atoms. These forms of matter are not regular lattices, but they can occur naturally. So Penrose's discovery changed our ideas about natural arrangements of atoms in crystal-like structures.

The detailed mathematics and crystallography are too complicated to describe here. To find out more, go to: en.wikipedia.org/wiki/Penrose_tiling

Answers on page 278

. . . .

Chaos Theory

If you want your friends to accept you as a 'rocket scientist', you have to be able to spout about *chaos theory*. You will casually mention the butterfly effect, and then you get to talk about where Pluto (no longer a planet but a mere dwarf planet) will be in 200 million years' time, and how really good dishwashers work.

Chaos theory is the name given by the media to an important new discovery in dynamical systems theory – the mathematics of systems that change over time according to specific rules. The name refers to a surprising and rather counterintuitive type of behaviour known as deterministic chaos. A system is called *deterministic* if its present state completely determines its future behaviour; if not, the system is called *stochastic* or *random*. Deterministic chaos – universally shortened to 'chaos' – is apparently random behaviour in a deterministic dynamical system. At first sight, this seems to be a contradiction in terms, but the issues are quite subtle, and it turns out that some features of deterministic systems can behave randomly.

Let me explain why.

You may remember the bit in Douglas Adams's *The Hitch Hiker's Guide to the Galaxy* that parodies the concept of determinism. No, really, you do – remember the supercomputer Deep Thought? When asked for the answer to the Great Question of Life, the Universe and Everything, it ruminates for five million years, and finally delivers the answer as 42. The philosophers then realise that they didn't actually understand the *question*, and an even greater computer is given the task of finding it.

Deep Thought is the literary embodiment of a 'vast intellect' envisaged by one of the great French mathematicians of the eighteenth century, the Marquis de Laplace. He observed that the laws of nature, as expressed mathematically by Isaac Newton and his successors, are deterministic, saying that: 'An intellect which at a certain moment knew all forces that set nature in motion, and all positions of all items of which nature is composed, if this intellect were also vast enough to submit these data to analysis, it would embrace in a single formula the movements of the greatest bodies of the universe and those of the tiniest atom; for such an intellect nothing would be uncertain and the future just like the past would be present before its eyes.'

In effect, Laplace was telling us that any deterministic system is inherently *predictable* – in principle, at least. In practice, however, we have no access to a Vast Intellect of the kind he had in mind, so we can't carry out the calculations that are needed to predict the system's future. Well, maybe for a short period, if we're lucky. For example, modern weather forecasts are fairly accurate for about two days, but a ten-day forecast is often badly wrong. (When it isn't, they've got lucky.)

Chaos raises another objection to Laplace's vision: even if his Vast Intellect existed, it would have to know 'all positions of all items' with *perfect* accuracy. In a chaotic system, any uncertainty about the present state grows very rapidly as time passes. So we quickly lose track of what the system will be doing. Even if this initial uncertainty first shows up in the millionth decimal place of some measurement – with the previous 999,999 decimal places absolutely correct – the predicted future based on one value for that millionth decimal place will be utterly different from a prediction based on some other value.

In a non-chaotic system such uncertainties grow fairly slowly, and very long-term predictions can be made. In a chaotic system, inevitable errors in measuring its state *now* mean that its state a short time ahead may be completely uncertain.

A (slightly artificial) example may help to clarify this effect. Suppose that the state of some system is represented by a real number – an infinite decimal – between 0 and 10. Perhaps its current value is 5.430 874, say. To keep the maths simple, suppose that time passes in discrete intervals – 1, 2, 3, and so on. Let's call these intervals 'seconds' for definiteness. Further, suppose that the rule for the future behaviour is this: to find the 'next' state – the state one second into the future – you take the current state, multiply by 10, and ignore any initial digit that would make the result bigger than 10. So the current value 5.430 874 becomes 54.308 74, and you ignore the initial digit 5 to get the next state, 4.308 74. Then, as time ticks on, successive states are:

> 5.430 874
> 4.308 74
> 3.0874
> 0.874
> 8.74
> 7.4

and so on.

Now suppose that the initial measurement was slightly inaccurate, and should have been 5.430 824 – differing in the fifth decimal place. In most practical circumstances, this is a very tiny error. Now the predicted behaviour would be:

> 5.430**8**24
> 4.30**8**24
> 3.0**8**24
> 0.**8**24
> **2**.4

See how that **2** moves one step to the left at each step – making the error ten times as big each time. After a mere 5 seconds, the first prediction of 7.4 has changed to 2.4 – a significant difference.

If we had started with a million-digit number, and changed the final digit, it would have taken a million seconds for the change to affect the predicted *first* digit. But a million seconds is only $11\frac{1}{2}$ days. And most mathematical schemes for predicting the future behaviour of a system work with much smaller intervals of time – thousandths or millionths of seconds.

If the rule for moving one time-step into the future is different, this kind of error may not grow as quickly. For example, if the rule is 'divide the number by 2', then the effect of such a change dies away as we move further and further into the future. So what makes a system chaotic, or not, is the *rule* for forecasting its next state. Some rules exaggerate errors, some filter them out.

The first person to realise that sometimes the error can grow rapidly – that the system may be chaotic, despite being deterministic – was Henri Poincaré, in 1887. He was competing for a major mathematical prize. King Oscar II of Norway and Sweden offered 2,500 crowns to anyone who could calculate whether the solar system is stable. If we wait long enough, will the planets continue to follow roughly their present orbits, or could something dramatic happen – such as two of them colliding, or one being flung away into the depths of interstellar space?

This problem turned out to be far too difficult, but Poincaré managed to make progress on a simpler question – a hypothetical solar system with just three bodies. The mathematics, even in this simplified set-up, was still extraordinarily difficult. But Poincaré was up to the task, and he convinced himself that this 'three-body' system sometimes behaved in an irregular, unpredictable manner. The equations were deterministic, but their solutions were erratic.

He wasn't sure what to do about that, but he knew it must be true. He wrote up his work, and won the prize.

Complicated orbits for three bodies moving under gravity.

And that was what everyone thought until recently. But in 1999 the historian June Barrow-Green discovered a skeleton in Poincaré's closet. The published version of his prizewinning paper was not the one he submitted, not the one that won the prize. The version he submitted – which was printed in a major mathematical journal – claimed that *no* irregular behaviour would occur. Which is the exact opposite of the standard story.

Barrow-Green discovered that shortly after winning the prize, an embarrassed Poincaré realised he had blundered. He withdrew the winning memoir and paid for the entire print run of the journal to be destroyed. Then he put his error right, and the official published version is the corrected one. No one knew that there had been a previous version until Barrow-Green discovered a copy tucked away among the archives of the Mittag-Leffler Institute in Stockholm.

Anyway, Poincaré deserves full credit as the first person to appreciate that deterministic mathematical laws do not always imply predictable, regular behaviour. Another famous advance was made by the meteorologist Edward Lorenz in 1961. He was running a mathematical model of convection currents on his computer. The machines available in those days were very slow and cumbersome compared with what we have now – your mobile phone is a far more powerful computer than the top research machine of the 1960s. Lorenz had to stop his computer in the middle of a long calculation, so he printed out all the numbers it had found. Then he went back several steps, input the numbers at that point, and restarted the calculation. The reason for backtracking was to check that the new calculation agreed with the old one, to eliminate errors when he fed the old figures back in.

It didn't.

At first the new numbers were the same as the old ones, but then they started to differ. What was wrong? Eventually Lorenz discovered that he hadn't typed in any wrong numbers. The difference arose because the computer stored numbers to a few more decimal places than it printed out. So what it stored as 2.371 45, say, was printed out as 2.371. When he typed that number in for the second run, the computer began calculating using 2.371 00, not 2.371 45. The difference grew – chaotically – and eventually became obvious.

When Lorenz published his results, he wrote: 'One meteorologist remarked that if the theory were correct, one flap of a

(Left) Initial conditions for eight weather forecasts, apparently identical but with tiny differences. (Right) The predicted weather a week later – the initial differences have grown enormously. Italian weather is more predictable than British. [Courtesy of the European Medium Range Weather Forecasting Centre, Reading.]

seagull's wings could change the course of weather for ever.' The objection was intended as a put-down, but we now know that this is exactly what happens. Weather forecasters routinely make a whole 'ensemble' of predictions, with slightly different initial conditions, and then take a majority vote on the future, so to speak.

Before you rush out with a shotgun, I must add that there are billions of seagulls, and we don't get to run the weather twice. What we end up with is a random selection from the range of possible weathers that might have happened instead.

Lorenz quickly replaced the seagull by a butterfly, because that sounded better. In 1972 he gave a lecture with the title 'Does the flap of a butterfly's wings in Brazil set off a tornado in Texas?' The title was invented by Philip Merilees when Lorenz failed to provide one. Thanks to this lecture, the mathematical point concerned became known as the butterfly effect. It is a characteristic feature of chaotic systems, and it is why they are unpredictable, despite being deterministic. The slightest change to the current state of the system can grow so rapidly that it changes the future behaviour. Beyond some relatively small 'prediction horizon', the future must remain mysterious. It may be predetermined, but we can't find out what has been predetermined, except by waiting to see what happens. Even a big increase in computer speed makes little difference to this horizon, because the errors grow so fast.

For weather, the prediction horizon is about two days ahead. For the solar system as a whole, it is far longer. We can predict that in 200 million years' time, Pluto will still be in much the same orbit as it is today; however, we have no idea on which side of the Sun it will be by then. So some features are predictable, others are not.

Although chaos is unpredictable, it is not random. This is the whole point. There are hidden 'patterns', but you have to know how to find them. If you plot the solutions of Lorenz's model in three dimensions, they form a beautiful, complicated shape called a strange attractor. If you plotted random data that way, you'd just get a fuzzy mess.

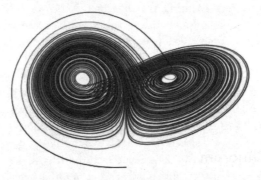

The Lorenz attractor, a geometric representation of Lorenz's calculations.

Chaos may seem a useless phenomenon, on the grounds that it prevents practical predictions. Even is this objection were correct, chaos would still exist. The real world is not obliged to behave in ways that are convenient for humans. As it happens, there are ways to make use of chaos. For a time, a Japanese company marketed a chaotic dishwasher, with two rotary arms spraying water on its contents. The resulting irregular spray cleaned the dishes better than the regular spray from a single rotating arm would have done.

And, of course, a dishwasher based on chaos theory was obviously very scientific and advanced. The marketing people must have loved it.

• •

Après-le-Ski

The little-known Alpine village of Après-le-Ski is situated in a deep mountain valley with vertical cliffs on both sides. The cliffs are 600 metres high on one side and 400 metres high on the other. A cable car runs from the foot of each cliff to the top of the other cliff, and the cables are perfectly straight. At what height above the ground do the two cables cross?

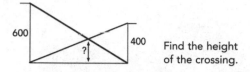

600   400   Find the height
      ?↓   of the crossing.

Answer on page 278

• •

Pick's Theorem

Here is a *lattice polygon*: a polygon whose vertices lie on the points of a square lattice. Assuming that the points are spaced at intervals of one unit, what is the area of the polygon?

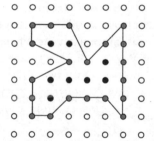

A lattice polygon.

There's a wonderfully simple way to find such areas, however complicated the polygon may be – by using *Pick's theorem*. It was proved by Georg Pick in 1899. For any lattice polygon, the area A can be calculated from the number of boundary points B (grey) and interior points I (black) by the formula

$$A = \tfrac{1}{2}B + I - 1$$

Here $B = 20$ and $I = 8$, so the area is $\tfrac{1}{2} \times 20 + 8 - 1 = 17$ square units.

What is the area of the lattice polygon in the second diagram?

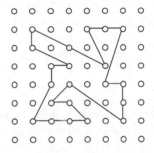

Find the area.

Answer on page 279

• •

Mathematical Prizes

There is no Nobel Prize in mathematics, but there are several equally prestigious prizes and a vast range of smaller ones, among them:

Fields Medal

The Fields Medal was instituted by the Canadian mathematician John Charles Fields and was first awarded in 1936. Every four years the International Mathematical Union selects for the award up to four of the world's leading research mathematicians, who must be under 40 years old. The prize consists of a gold medal and a small sum of money – currently around $13,500 – but is considered equivalent to a Nobel Prize in prestige.

Abel Prize

In 2001 the Norwegian government commemorated the 200th anniversary of the birth of Niels Henrik Abel – one of the all-time greats of mathematics – with a new prize. Each year, one or more mathematicians share a prize in the region of $1,000,000, which is comparable to the sum that Nobel Prize winners receive. The King of Norway presents the award at a special ceremony.

Shaw Prize

Sir Run Run Shaw, a prominent figure in Hong Kong's media and a long-standing philanthropist, established an annual prize for three areas of science: astronomy, life sciences and medicine, and mathematics. The total value awarded each year is $1,000,000, and there is also a medal. The first Shaw Prize was awarded in 2002.

Clay Millennium Prizes

The Clay Mathematics Institute in Cambridge, Massachusetts, founded by Boston businessman Landon T. Clay and Lavinia D. Clay, offers seven prizes, each of $1,000,000, for the definitive solution of seven major open problems. These 'Millennium Prize Problems' were selected to represent some of the biggest challenges facing mathematicians. For the record, they are:

- The Birch and Swinnerton-Dyer Conjecture in algebraic number theory.
- The Hodge Conjecture in algebraic geometry.
- The existence of solutions, valid for all time, to the Navier–Stokes equations of fluid dynamics.
- The P = NP? problem in computer science
- The Poincaré Conjecture in topology.
- The Riemann Hypothesis in complex analysis and the theory of prime numbers.
- The mass gap hypothesis and associated issues for the Yang–Mills equations in quantum field theory.

None of the prizes has yet been awarded, but the Poincaré Conjecture has now been proved. The main breakthrough was made by Grigori Perelman, and many details were clarified by other mathematicians. For details of the seven problems, see www.claymath.org/millennium/

King Faisal International Prize

Between 1977 and 1982 the King Faisal Foundation instituted prizes for service to Islam, Islamic studies, Arabic literature, medicine and science. The science prize is open to, and has been won by, mathematicians. The winner receives a certificate, a gold medal and SR 750 000 ($200,000).

Wolf Prize

Since 1978 this prize has been awarded by the Wolf Foundation, set up by Ricardo Wolf and his wife Francisca Subirana Wolf. It covers five areas of science: agriculture, chemistry, mathematics, medicine and physics. The prize consists of a diploma and $100,000.

Beal Prize

In 1993 Andrew Beal, a Texan with a passion for number theory, was led to conjecture that if $a^p + b^q = c^r$, where a, b, c, p, q and r are positive integers, and p, q and r are all greater than 2, then a, b and c must have a common factor. In 1997 he offered a prize, later increased to $100,000, for a proof or disproof.

. . . .

Why No Nobel for Maths?

Why didn't Alfred Nobel set up a mathematics prize? There's a persistent story that Nobel's wife had an affair with the Swedish mathematician Gosta Mittag-Leffler, so Nobel hated mathematicians. But there's a problem with this theory, because Nobel never married. Some versions of the story replace the hypothetical wife with a fiancée or a mistress. Nobel may have had a mistress – a Viennese lady called Sophie Hess – but there's no evidence that she had anything to do with Mittag-Leffler.

An alternative theory holds that Mittag-Leffler, who became quite wealthy himself, did something to annoy Nobel. Since Mittag-Leffler was the leading Swedish mathematician of the time, Nobel realised that he was very likely to win a prize for mathematics, and decided not to set one up. However, in 1985 Lars Gårding and Lars Hörmander noted that Nobel left Sweden in 1865, to live in Paris, and seldom returned – and in 1865 Mittag-Leffler was a young student. So there was little opportunity for them to interact, which casts doubt on both theories.

It's true that late in Nobel's life, Mittag-Leffler was chosen to negotiate with him about leaving to the Stockholm Högskola (which later became the University) a significant amount of money in his will, and this attempt eventually failed – but presumably Mittag-Leffler wouldn't have been chosen if he'd already offended Nobel. In any case, Mittag-Leffler wasn't likely to win a mathematical Nobel if one existed – there were plenty of more prominent mathematicians around. So it seems more likely that it simply never occurred to Nobel to award a prize for mathematics, or that he considered the idea and rejected it, or that he didn't want to spend even more cash.

Despite this, several mathematicians and mathematical physicists have won the prize for work in other areas – physics, chemistry, physiology/medicine, even literature. They have also won the 'Nobel' in economics – the Prize in Economic Sciences in Memory of Alfred Nobel, established by the Sveriges Riksbank in 1968.

• •

Is There a Perfect Cuboid?

It is easy to find rectangles whose sides and diagonals are whole numbers – this is the hoary old problem of Pythagorean triangles, and it has been known since antiquity how to find all of them (page 58). Using the classical recipe, it is not too hard to find a cuboid – a box with rectangular sides – such that its sides, and the diagonals of all its faces, are whole numbers. The first set of values given below achieves this. But what no one has yet been able to find is a *perfect* cuboid – one in which the 'long diagonal' between opposite corners of the cuboid is also a whole number.

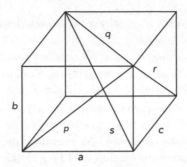

Make all lengths
integers.

With the notation in the diagram, and bearing Pythagoras in mind, we have to find a, b and c so that all four of the numbers $a^2 + b^2, a^2 + c^2, b^2 + c^2$ and $a^2 + b^2 + c^2$ are perfect squares – equal, respectively, to p^2, q^2, r^2 and s^2. The existence of such numbers has neither been proved nor disproved, but some 'near misses' have been found:

$a = 240$, $b = 117$, $c = 44$, $p = 267$, $q = 244$, $r = 125$,
but s is not an integer
$a = 672$, $b = 153$, $c = 104$, $q = 680$, $r = 185$, $s = 697$,
but p is not an integer
$a = 18{,}720$, $b = 211{,}773{,}121$, $c = 7{,}800$, $p = 23{,}711$,
$q = 20{,}280$, $r = 16{,}511$, $s = 24{,}961$, but b is not an integer

If there is a perfect cuboid, it involves big numbers: it has been proved that the smallest edge is at least $2^{32} = 4{,}294{,}967{,}296$.

•••

Paradox Lost

In mathematical logic, a *paradox* is a self-contradictory statement – the best known is 'This sentence is a lie.' Another is Bertrand Russell's 'barber paradox'. In a village there is a barber who shaves everyone who does not shave themselves. So who shaves the barber? Neither 'the barber' nor 'someone else' is logically acceptable. If it is the barber, then he shaves himself – but we are told that he doesn't. But if it's someone else, then the barber does not shave himself ... but we are told that he shaves all such people, so he does shave himself.

In the real world, there are plenty of get-outs (are we talking about shaving beards here, or legs, or what? Is the barber a woman? Can such a barber actually exist anyway?) But in mathematics, a more carefully stated version of Russell's paradox ruined the life's work of Gottlob Frege, who attempted to base the whole of mathematics on set theory – the study of collections of objects, and how these can be combined to form other collections.

Here's another famous (alleged) paradox:

Protagoras was a Greek lawyer who lived and taught in the fifth century BC. He had a student, and it was agreed that the student would pay for his teaching after he had won his first case. But the student didn't get any clients, and eventually Protagoras threatened to sue him. Protagoras reckoned that he would win either way: if the court upheld his case, the student would be required to pay up, but if Protagoras lost, then by their agreement the student would have to pay anyway. The student argued exactly the other way round: if Protagoras won, then by their agreement the student did *not* have to pay, but if Protagoras lost, the court would have ruled that the student did not have to pay.

Is this a genuine logical paradox or not?

Answer on page 279

Answer on page 279

• •

When Will My MP3 Player Repeat?

You have 1,000 songs on your MP3 player. If it plays songs 'at random', how long would you expect to wait before the same song is repeated?

It all depends on what 'at random' means. The leading MP3 player on the market 'shuffles' songs just like someone shuffling a pack of cards. Once the list has been shuffled, all songs are played in order. If you don't reshuffle, it will take 1,001 songs to get a repeat. However, it is also possible to pick a song at random, and keep repeating this procedure without eliminating that song. If so, the same song might – just – come up twice in a row. I'll assume that all songs appear with the same probability, though some MP3 players bias the choice in favour of songs you play a lot.

You've probably met the same problem with birthdays replacing songs. If you ask people their birthday, one at a time, then on average how many do you have to ask to get a repeat? The answer is 23, remarkably small. There is a second, super-ficially similar problem: how many people should there be at a party so that the probability that at least two share a birthday is bigger than $\frac{1}{2}$? Again the answer is 23. In both calculations we ignore leap years and assume that any particular birthday occurs with probability 1/365. This isn't quite accurate, but it simplifies the sums. We also assume that all individuals have statistically independent birthdays, which would not be the case if, say, the party included twins.

I'll solve the second birthday problem, because the sums are easier to understand. The trick is to imagine the people entering the room one at a time, and to work out, at each stage, the probability that all birthdays so far are *different*. Subtract the result from 1 and you get the probability that at least two are equal. So we want to continue allowing people to enter until the probability that all birthdays are different drops *below* $\frac{1}{2}$.

When the first person enters, the probability that their birthday is different from that of anyone else present is 1, because nobody else is present. I'll write that as the fraction

$$\frac{365}{365}$$

because it tells us that out of the 365 possible birthdays, all 365 have the required outcome.

When the second person enters, their birthday has to be

different, so there are now only 364 choices out of 365. So the probability we want is now

$$\frac{365}{365} \times \frac{364}{365}$$

When the third person enters, they have only 363 choices, and the probability of no duplication so far is

$$\frac{365}{365} \times \frac{364}{365} \times \frac{363}{365}$$

The pattern should now be clear. After k people have entered, the probability that all k birthdays are distinct is

$$\frac{365}{365} \times \frac{364}{365} \times \frac{363}{365} \times \cdots \times \frac{365 - k + 1}{365}$$

and we want the first k for which this is less than $\frac{1}{2}$. Each fraction, other than the first, is smaller than 1, so the probability decreases as k increases. Direct calculation shows that when $k = 22$ the fraction equals 0.524 305, and when k is 23 it equals 0.492 703. So the required number of people is 23.

This number seems surprisingly small, which may be because we confuse the question with a different one: how many people do you have to ask for the probability that one of them has the same birthday as *you* to be bigger than $\frac{1}{2}$? The answer to that is much bigger – in fact, it's 253.

The same calculation with 1,000 songs on an MP3 player shows that if each song is chosen at random then you have to play a mere 38 songs to make the probability of a repeat bigger than $\frac{1}{2}$. The *average* number of songs you have to play to get a repeat is 39 – slightly more.

These sums are all very well, but they don't provide much insight. What if you had a million songs? It's a big sum – a computer can do it, though. But is there a simpler answer? We can't expect an exact formula, but we ought to be able to find a good approximation. Let's say we have n songs. Then it turns out that on average we have to play approximately

$$\sqrt{(\tfrac{1}{2}\pi)}\sqrt{n} = 1.2533\sqrt{n}$$

songs to get a repeat (of *some* song already played, not necessarily the first one). To make the probability of a repeat greater than $\frac{1}{2}$, we have to play approximately

$$\sqrt{(\log 4)}\sqrt{n}$$

songs, which is

$$1.1774\sqrt{n}$$

This is about 6% smaller.

Both numbers are proportional to the square root of n, which grows much more slowly than n. This is why we get quite small answers when n is large. If you did have a million songs on your MP3 player, then on average you would have to play only 1,253 of them to get a repeat (the square root of a million is 1,000). And to make the probability of a repeat greater than $\frac{1}{2}$, you would have to play approximately 1,177 of them. The exact answer, according to my computer, is 1,178.

● ●

Six Pens

Farmer Hogswill has run into another mathematico-agricultural problem. He had carefully assembled 13 identical fence panels to create 6 identical pens for his rare-breed Alexander-horned pigs. But during the night some antisocial person stole one of his panels. So now he needs to use 12 fence panels to create 6 identical pens. How can he achieve this? All 12 panels must be used.

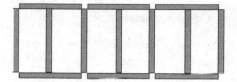

13 panels
making 6 pens.

Answer on page 280

● ●

Patented Primes

Because of their importance in encryption algorithms, prime
numbers have commercial significance. In 1994 Roger Schlafly
obtained US Patent 5,373,560 on two primes. The patent states
them as hexadecimal (base-16) numbers, but I've converted
them into decimal. They are:

> 7,994,412,097,716,110,548,127,211,733,331,600,522,933,
> 776,757,046,707,649,963,673,962,686,200,838,432,950,239,
> 103,981,070,728,369,599,816,314,646,482,720,706,826,018,
> 360,181,196,843,154,224,748,382,211,019

and

> 103,864,912,054,654,272,074,839,999,186,936,834,171,066,
> 194,620,139,675,036,534,769,616,693,904,589,884,931,513,
> 925,858,861,749,077,079,643,532,169,815,633,834,450,952,
> 832,125,258,174,795,234,553,238,258,030,222,937,772,878,
> 346,831,083,983,624,739,712,536,721,932,666,180,751,292,
> 001,388,772,039,413,446,493,758,317,344,413,531,957,900,
> 028,443,184,983,069,698,882,035,800,332,668,237,985,846,
> 170,997,572,388,089

He did this to publicise deficiencies in the US patent system.

Legally, you can't use these numbers without Schlafly's
permission. Hmmm ...

● ●

The Poincaré Conjecture

Towards the end of the nineteenth century, mathematicians succeeded in finding all possible 'topological types' of surfaces. Two surfaces have the same topological type if one of them can be continuously deformed into the other. Imagine that the surface is made from flexible dough. You can stretch it, squeeze it, twist it – but you can't tear it, or squash different bits together.

To keep the story simple, I'll assume that the surface has no boundary, that it's orientable (two-sided, unlike the Möbius band) and that it's of finite extent. The nineteenth-century mathematicians proved that every such surface is topologically equivalent to a sphere, a torus, a torus with two holes, a torus with three holes, and so on.

Sphere.   Torus.   Two-holed torus.

'Surface' here really does refer only to the *surface*. A topologist's sphere is like a balloon – an infinitely thin sheet of rubber. A torus is shaped like an inner tube for a tyre (for those of you who know what an inner tube is). So the 'dough' I just mentioned is really a very thin sheet, not a solid lump. Topologists call a solid sphere a 'ball'.

To achieve their classification of all surfaces, the topologists had to characterise them 'intrinsically', without reference to any surrounding space. Think of an ant living on the surface, ignorant of any surrounding space. How can it work out which surface it inhabits? By 1900 it was understood that a good way to answer such questions is to think about closed loops in the surface, and how these loops can be deformed. For example, on a sphere (by which I mean just the surface, not the solid interior) any closed loop can be continuously deformed to a point – 'shrunk'. For example, the circle running round the equator can be gradually moved towards the south pole, becoming smaller and smaller until it coincides with the pole itself:

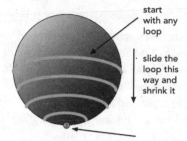

start with any loop

slide the loop this way and shrink it

How to shrink a loop on a sphere continuously to a point.

In contrast, every surface that is not equivalent to a sphere contains loops, which cannot be deformed to points. Such loops 'pass through a hole', and the hole prevents them from being shrunk. So the sphere can be characterised as the *only* surface on which any closed loop can be shrunk to a point.

Observe, however, that the 'hole' that we see in a picture is not actually part of the surface. By definition, it's a place where the surface *isn't*. If we think intrinsically, we can't talk sensibly about holes if we try to visualise them in the usual manner. The ant who lives on the surface and knows no other universe can't see that his torus has a dirty great hole in it – any more than we can look along a fourth dimension. So although I'm using the word 'hole' to explain why the loop can't be shrunk, a topological proof runs along different lines.

a loop like this one can't be shrunk

On all other surfaces, loops can get stuck.

In 1904 Henri Poincaré was trying to take the next step, and understand 'manifolds'– three-dimensional analogues of surfaces – and for a time he assumed that the characterisation of a sphere in terms of shrinking loops is also true in three dimensions, where there is a natural analogue of the sphere called the 3-sphere. A 3-sphere is *not* just a solid ball, but it can be visualised – if that's the word – by taking a solid ball and pretending that its entire surface is actually just a single point.

Imagine doing the same with a circular disc. The rim closes up like the top of a bag as you draw a string tight round its edge, and the result is topologically a sphere. Now go up a dimension ...

take a disc...

...scrunge the edge together...

...and you get a sphere

Turning a disc into a sphere.

At first, Poincaré thought that this characterisation of the 3-sphere should be obvious, or at least easily proved, but later he realised that one plausible version of this statement is actually wrong, while another closely related formulation seemed difficult to prove but might well be true. He posed a deceptively simple question: if a three-dimensional manifold (without boundary, of finite extent, and so on) had the property that any loop in it can be shrunk to a point, must that manifold be topologically equivalent to the 3-sphere?

Subsequent attempts to answer this question failed dismally, although after a huge effort by the world's topologists, the answer has proved to be 'yes' for all versions in every dimension *higher* than 3. The belief that the same answer applies in three dimensions became known as the *Poincaré Conjecture*, famous as one of the eight Millennium Prize Problems (page 127).

In 2002 a Russian-born mathematician, Grigori Perelman, caused a sensation by placing several papers on arXiv.org, an informal website for current research in physics and mathematics. His papers were ostensibly about various properties of the 'Ricci flow', but it became clear that if the work was correct, it implied that the Poincaré Conjecture is also correct. The idea of using the Ricci flow dates to 1982, when Richard Hamilton introduced a new technique based on mathematical ideas used by Albert Einstein in general relativity. According to Einstein, spacetime can be considered as curved, and the curvature describes the force of gravity. Curvature is measured by something called the 'curvature tensor', and this has a simpler relative known as the 'Ricci tensor' after its inventor, Gregorio Ricci-Curbastro.

According to general relativity, gravitational fields can change the geometry of the universe as time passes, and these changes are governed by the Einstein equations, which say that the stress tensor is proportional to the curvature. In effect, the gravitational bending of the universe tries to smooth itself out as time passes, and the Einstein equations quantify that idea.

The same game can be played using the Ricci version of curvature, and it leads to the same kind of behaviour: a surface that obeys the equations for the Ricci flow will naturally tend to simplify its own geometry by redistributing its curvature more evenly. Hamilton showed that the familiar two-dimensional version of the Poincaré Conjecture, characterising the sphere, can be proved using the Ricci flow. Basically, a surface in which all loops shrink simplifies itself so much as it follows the Ricci flow that it ends up being a perfect sphere. Hamilton suggested generalising this approach to three dimensions, but he hit some difficult obstacles.

The main complication in three dimensions is that 'singularities' can develop, where the manifold pinches together and the flow breaks down. Perelman's new idea was to cut the surface apart near such a singularity, cap off the resulting holes, and then allow the flow to continue. If the manifold manages to simplify itself completely after only finitely many singularities have arisen, then not only is the Poincaré Conjecture true, but a more far-reaching result, the Thurston Geometrisation Conjecture, is also true. And that tells us about *all possible* three-dimensional manifolds.

Now the story takes a curious turn. It is generally accepted that Perelman's work is correct, although his arXiv papers leave a lot of gaps that have to be filled in correctly, and that has turned out to be quite difficult. Perelman had his own reasons for not wanting the prize – indeed, any reward save the solution itself – and decided not to expand his papers into something suitable for publication, although he was generally willing to explain how to fill in various details if anyone asked him. Experts in the area were forced to develop their own versions of his ideas.

Perelman was also awarded a Fields Medal at the Madrid International Congress of Mathematicians in 2006, the top prize in mathematics. He turned that down, too.

Hippopotamian Logic

> I won't eat my hat.
> If hippos don't eat acorns, then oak trees will grow in Africa.
> If oak trees don't grow in Africa, then squirrels hibernate in winter.
> If hippos eat acorns and squirrels hibernate in winter, then I'll eat my hat.
> Therefore – *what*?

Answer on page 280

Langton's Ant

Langton's ant was invented by Christopher Langton, and it shows how amazingly complex simple ideas can be. It leads to one of the most baffling unsolved problems in the whole of mathematics, and all from astonishingly simple ingredients.

The ant lives on an infinite square grid of black and white cells, and it can face in one of the four compass directions: north, south, east or west. At each tick of a clock it moves one cell forward, and then follows three simple rules:

- If it lands on a black cell it makes a 90° turn to the left.
- If it lands on a white cell it makes a 90° turn to the right.
- The cell that it has just vacated then changes colour, from white to black, or vice versa.

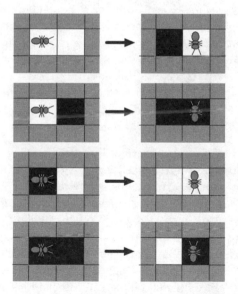

Effect of the ant moving. Grey cells can be any colour and do not change on this move.

As a warm-up, the ant starts by facing east on a completely white grid. Its first move takes it to a white square, while the square it started from turns black. Because it is on a white square, the ant's next move is a right turn, so now it faces south. That takes it to a new white square, and the square it has just vacated turns black. After a few more moves the ant starts to revisit earlier squares that have turned black, so it then turns to the left instead. As time passes, the ant's motion gets quite complicated, and so does the ever-changing pattern of black and white squares that trails behind it.

Jim Propp discovered that the first few hundred moves occasionally produce a nice, symmetrical pattern. Then things get rather chaotic for about ten thousand moves. After that, the ant gets trapped in a cycle in which the same sequence of 104 moves is repeated indefinitely, each cycle moving it two squares diagonally. It continues like this for ever, systematically building a broad diagonal 'highway'.

Langton's ant builds a highway.

This 'order out of chaos' behaviour is already puzzling, but computer experiments suggest something more surprising. If you scatter any finite number of black squares on the grid, before the ant sets off, it *still* ends up building a highway. It may take longer to do so, and its initial movements may be very different, but ultimately that's what will happen. As an example, the second diagram shows a pattern that forms when the ant starts inside a solid rectangle. Before building its highway, the ant builds a 'castle' with straight walls and complicated crenellations. It keeps destroying and rebuilding these structures in a curiously purposeful way, until it gets distracted and wanders off – building a highway.

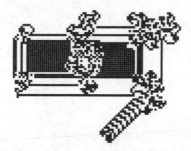

Pattern created by Langton's ant when it starts inside a black rectangle. The highway is at the lower right. Small white dots mark squares of the original rectangle that have never been visited.

The problem that is baffling mathematicians is to prove that the ant *always* ends up building a highway, for every initial configuration of finitely many black squares. Or disprove that, if it's wrong. We do know that the ant can never get trapped inside any bounded region of the grid – it always escapes if you wait long enough. But we don't know that it escapes along a highway.

• •

Pig on a Rope

Farmer Hogswill owns a field, which is a perfect equilateral triangle, each side 100 metres long. His prize pig Pigasus is tied to one corner, so that the portion of the field that Pigasus can reach is exactly half the total area. How long is the rope?

You may – indeed, must – assume that the pig has zero size (which admittedly is pretty silly) and that the rope is infinitely thin and any necessary knots can be ignored.

Pigs may safely graze ... over half the area of the field.

Answer on page 280

• •

The Surprise Examination

This paradox is so famous that I nearly left it out. It raises some intriguing issues.

Teacher tells the class that there will be a test one day next week (Monday to Friday), and that it will be a surprise. This seems reasonable: the teacher can choose any day out of five, and there is no way that the students can know which day it will be. But the students don't see things that way at all. They reason that the test can't be on Friday – because if it was, then as soon as Thursday passed without a test, they'd know it had to be Friday, so no surprise. And once they've ruled out Friday, they apply the same reasoning to the remaining four days of the week, so the test can't be on Thursday, either. In which case it can't be on Wednesday, so it can't be on Tuesday, so it can't be on Monday. Apparently, no surprise test is possible.

That's all very well, but if the teacher decides to set the test on Wednesday, there seems to be no way that the students could actually *know* the day ahead of time! Is this a genuine paradox or not?

Answer on page 281

• •

Antigravity Cone

In defiance of the Law of Gravity, this double cone *rolls uphill*. Here's how to make it.

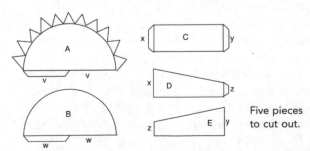

Five pieces to cut out.

Copy the five shapes on to a thin sheet of card, two or three times the size shown here, and cut them out. On piece A, glue flap v to edge v to make a cone. On piece B, glue flap w to edge w to make a second cone. Then glue the two cones base to base using the triangular flaps on A.

Glue flap x of C to edge x of D, and flap y of C to edge y of E. Finally, glue flap z of D to edge z of E to make a triangular 'fence'.

Place the double cone at the lower end of this triangle, and let go. It will appear to roll uphill.

How can this happen?

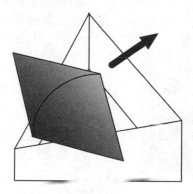

A rolling cone gathers no gravity.

Answer on page 282

Mathematical Jokes 2

An engineer, a physicist, and a mathematician are staying in a hotel. The engineer wakes up and smells smoke. He goes into the hallway, sees a fire, fills the wastepaper basket from his room with water, and pours it on the fire, putting it out.

Later, the physicist wakes up and smells smoke. He goes into the hallway and sees a (second) fire. He pulls a fire hose off the wall. Having calculated the temperature of the exothermic reaction, the velocity of the flame front, the water pressure in the hose, and so on, he uses the hose to put out the fire with the minimum expenditure of energy.

Later, the mathematician wakes up and smells smoke. He goes into the hallway and sees a (third) fire. He notices the fire hose on the wall, and thinks for a moment ... Then he says, 'OK, a solution exists!' – and goes back to bed.

●●

Why Gauss Became a Mathematician

Carl Friedrich Gauss.

Carl Friedrich Gauss was born in Brunswick in 1777 and died in Göttingen in 1855. His parents were uneducated manual workers, but he became one of the greatest mathematicians ever; many consider him the best. He was precocious – he is said to have pointed out a mistake in his father's financial calculations when he was three. At the age of nineteen he had to decide whether to study mathematics or languages, and the decision was made for him when he discovered how to construct a regular 17-sided polygon using the traditional Euclidean tools of an unmarked ruler and a compass.

This may not sound like much, but it was totally unprecedented, and the discovery led to a new branch of number theory. Euclid's *Elements* contains constructions for regular polygons (all sides equal length, all angles equal) with 3, 4, 5, 6 and 15 sides, and the ancient Greeks knew that the number of sides could be doubled as often as you wish. Up to 100, the number of sides in a constructible (regular) polygon – as far as the Greeks knew – must be

2, 3, 4, 5, 6, 8, 10, 12, 15, 16, 20, 24, 30, 32, 40, 48, 60, 64, 80, 96

For more than two thousand years, everyone assumed that no other polygons were constructible. In particular, Euclid does not tell us how to construct 7-gons or 9-gons, and the reason is that he had no idea how this might be done. Gauss's discovery was a bombshell, adding 17, 34 and 68 to the list. Even more amazingly, his methods prove that other numbers, such as 7, 9, 11 and 13, are impossible. (The polygons do exist, but you can't construct them by Euclidean methods.)

Gauss's construction depends on two simple facts about the number 17: it is prime, and it is one greater than a power of 2. The whole problem pretty much reduces to finding which prime numbers correspond to constructible polygons, and powers of 2 come into the story because every Euclidean construction boils down to taking a series of square roots – which in particular implies that the lengths of any lines that feature in the construction must satisfy algebraic equations whose degree is a power of two. The key equation for the 17-gon is

$$x^{16} + x^{15} + x^{14} + x^{13} + x^{12} + x^{11} + x^{10} + x^9 + x^8 + x^7 + x^6 + x^5 + x^4 + x^3 + x^2 + x + 1 = 0$$

where x is a *complex* number. The 16 solutions, together with the number 1, form the vertices of a regular 17-gon in the complex plane. Since 16 is a power of 2, Gauss realised that he was in with a chance. He did some clever calculations, and proved that the

17-gon can be constructed provided you can construct a line whose length is

$$\frac{1}{16}\left[-1 + \sqrt{17} + \sqrt{34 - 2\sqrt{17}} + \right.$$

$$\left. \sqrt{68 + 12\sqrt{17} - 16\sqrt{34 + 2\sqrt{17}} - 2(1 - \sqrt{17})(\sqrt{34 - 2\sqrt{17}})} \right]$$

Since you can always construct square roots, this effectively solves the problem, and Gauss didn't bother to describe the precise steps needed – the formula itself does that. Later, other mathematicians wrote down explicit constructions. Ulrich von Huguenin published the first in 1803, and H.W. Richmond found a simpler one in 1893.

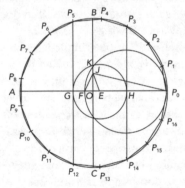

Richmond's method for constructing a regular 17-gon. Take two perpendicular radii, AOP_0 and BOC, of a circle. Make $\frac{OJ}{4OB} = 1$ and angle $\frac{OJE}{4OJP_0} = 1$. Find F Such that angle EJF is 45°. Draw a circle with FP_0 as diameter, meeting OB at K. Draw the circle with centre E through K, cutting AP_0 in G and H. Draw HP_3 and GP_5 perpendicular to AP_0. Then P_0, P_3 and P_5 are respectively the 0th, 3rd and 5th vertices of a regular 17-gon, and the other vertices are now easily constructed.

Gauss's method proves that a regular n-gon can be constructed whenever n is a prime of the form $2^k + 1$. Primes like this are called *Fermat primes*, because Fermat investigated them. In particular he noticed that k must itself be a power of 2 if $2^k + 1$ is going to be prime. The values $k = 1, 2, 4, 8$ and 16 yield the Fermat primes 3, 5, 17, 257 and 65,537. However, $2^{32} + 1 = 4,294,967,297 = 641 \times 6,700,417$ is not prime. Gauss was aware that the regular n-gon is constructible if and only if n is a power of 2, or a power of 2 multiplied by *distinct* Fermat primes. But he didn't give a complete proof – probably because to him it was obvious.

His results prove that it is impossible to construct regular 7-, 11- or 13-gons by Euclidean methods, because these are prime but not of Fermat type. The analogous equation for the 7-gon, for instance, is $x^6 + x^5 + x^4 + x^3 + x^2 + x + 1 = 0$, and that has degree 6, which is not a power of 2. The 9-gon is not constructible because 9 is not a product of distinct Fermat primes – it is 3×3, and 3 is a Fermat prime, but the same prime occurs twice here.

The Fermat primes just listed are the only *known* ones. If there is another, it must be absolutely gigantic: in the current state of knowledge the first candidate is $2^{33,554,432} + 1$, where $33,554,432 = 2^{25}$. Although we're still not sure exactly which regular polygons are constructible, the only obstacle is the possible existence of very large Fermat primes. A useful website for Fermat primes is mathworld.wolfram.com/FermatNumber.html

In 1832 Friedrich Julius Richelot published a construction for the regular 257-gon. Johann Gustav Hermes of Lingen University devoted ten years to the 65,537-gon, and his unpublished work can be found at the University of Göttingen, but it probably contains errors.

With more general construction techniques, other numbers are possible. If you use a gadget for trisecting angles, then the 9-gon is easy. The 7-gon turns out to be possible too, but that's nowhere near as obvious.

What Shape is a Crescent Moon?

The Moon is low in the sky shortly after sunset or before dawn; the bright part of its surface forms a beautiful crescent. The two curves that form the boundary of the crescent resemble arcs of circles, and are often drawn that way. Assuming the Moon to be a perfect sphere, and the Sun's rays to be parallel, *are* they arcs of circles?

A crescent formed by two arcs of circles. Is the crescent Moon like this?

Answer on page 283

• •

Famous Mathematicians

All the people listed below – except one – either started a degree (or joint degree) in mathematics, or studied under famous mathematicians, or were professional mathematicians in their other life. What are they famous for? Which person does not belong on the list?

Pierre Boulez

Sergey Brin

Lewis Carroll

J.M. Coetzee

Alberto Fujimori

Art Garfunkel

Philip Glass

Teri Hatcher

Edmund Husserl

Michael Jordan

Theodore Kaczynski

John Maynard Keynes

Carole King

Emanuel Lasker

J.P. Morgan

Larry Niven

Alexander Solzhenitsyn

Bram Stoker

Leon Trotsky   Virginia Wade
Eamon de Valera   Ludwig Wittgenstein
Carol Vorderman   Sir Christopher Wren

Answers on page 284

• •

What is a Mersenne Prime?

A *Mersenne number* is a number of the form $2^n - 1$. That is, it is
one less than a power of 2. A *Mersenne prime* is a Mersenne
number that happens also to be prime. It is straightforward to
prove that in this case the exponent n must itself be prime. For
the first few primes, $n = 2$, 3, 5 and 7, the corresponding
Mersenne numbers 3, 7, 31 and 127 are all prime.

Interest in Mersenne numbers goes back a long way, and
initially it was thought that they are prime whenever n is prime.
However, in 1536 Hudalricus Regius proved that this assumption
is false, pointing out that $2^{11} - 1 = 2,047 = 23 \times 89$. In 1603 Pietro
Cataldi noted that $2^{17} - 1$ and $2^{19} - 1$ are prime, which is correct,
and claimed that $n = 23$, 29, 31 and 37 also lead to primes.
Fermat proved that he was wrong for 23 and 37, and Euler
demolished his claim for 29. But Euler later proved that $2^{31} - 1$ is
prime.

In his 1644 book *Cogitata Physico-Mathematica*, the French
monk Marin Mersenne stated that $2^n - 1$ is prime when n is 2, 3,
5, 7, 13, 17, 19, 31, 67, 127 and 257 – and for no other values in
that range. Using the methods then available, he could not have
tested most of these numbers, so his claims were mainly
guesswork, but his name became associated with the problem.

In 1876 Édouard Lucas developed a cunning way to test
Mersenne numbers to see if they are prime, and showed that
Mersenne was right for $n = 127$. By 1947 all cases in Mersenne's
range had been checked, and it turned out that he had
mistakenly included 67 and 257. He had also omitted 61, 89 and

107. Lucas improved his test, and in the 1930s Derrick Lehmer found further improvements. The Lucas–Lehmer test uses the sequence of numbers

4, 14, 194, 37634, ...

in which each number is the square of the previous one, decreased by 2. It can be proved that the nth Mersenne number is prime if and only if it divides the $(n-1)$th term of this sequence. This test can prove that a Mersenne number is composite without finding any of its prime factors, and it can prove the number is prime without testing for any prime factors. There's a trick to keep all numbers involved in the test smaller than the Mersenne number concerned.

Looking for new, larger Mersenne primes is an amusing way to try out new, fast computers, and over the years prime-hunters have extended the list. It now includes 44 primes:

n	Year	Discoverer
2	—	known from antiquity
3	—	known from antiquity
5	—	known from antiquity
7	—	known from antiquity
13	1456	anonymous
17	1588	Pietro Cataldi
19	1588	Pietro Cataldi
31	1772	Leonhard Euler
61	1883	Ivan Pervushin
89	1911	R.E. Powers*
107	1914	R.E. Powers
127	1876	Édouard Lucas
521	1952	Raphael Robinson
607	1952	Raphael Robinson
1,279	1952	Raphael Robinson
2,203	1952	Raphael Robinson
2,281	1952	Raphael Robinson
3,217	1957	Hans Riesel

* Powers is a rather obscure, possibly amateur, mathematician. I haven't been able to locate his first name.

4,253	1961	Alexander Hurwitz
4,423	1961	Alexander Hurwitz
9,689	1963	Donald Gillies
9,941	1963	Donald Gillies
11,213	1963	Donald Gillies
19,937	1971	Bryant Tuckerman
21,701	1978	Landon Noll and Laura Nickel
23,209	1979	Landon Noll
44,497	1979	Harry Nelson and David Slowinski
86,243	1982	David Slowinski
110,503	1988	Walter Colquitt and Luther Welsh
132,049	1983	David Slowinski
216,091	1985	David Slowinski
756,839	1992	David Slowinski *et al.*
859,433	1994	David Slowinski and Paul Gage
1,257,787	1996	David Slowinski and Paul Gage
1,398,269	1996	Joel Armengaud *et al.*
2,976,221	1997	Gordon Spence *et al.*
3,021,377	1998	Roland Clarkson *et al.*
6,972,593	1999	Nayan Hajratwala *et al.*
13,466,917	2001	Michael Cameron *et al.*
20,996,011	2003	Michael Shafer *et al.*
24,036,583	2004	Josh Findley *et al.*
25,964,951	2005	Martin Nowak *et al.*
30,402,457	2005	Curtis Cooper *et al.*
32,582,657	2006	Curtis Cooper *et al.*
37,156,667	2008	Hans-Michael Elvenich
43,112,609	2008	Edson Smith

Up to and including the 39th Mersenne prime ($n = 13,466,917$) the list is complete, but there may be undiscovered Mersenne primes in the gaps between the known ones after that. The 46th known Mersenne prime, $2^{43,112,609} - 1$, has 12,978,189 decimal digits and is currently (November 2008) the largest known prime. Mersenne primes generally hold this record, thanks to the Lucas–Lehmer test; however, we know from Euclid that there is no largest prime. For up-to-date information, go to the Mersenne Primes website primes.utm.edu/mersenne/; you can also join the Great Internet Mersenne Prime Search (GIMPS) at www.mersenne.org/

The Goldbach Conjecture

In 2000, as a publicity stunt for Apostolos Doxiadis's novel *Uncle Petros and Goldbach's Conjecture*, the publisher Faber & Faber offered a million-dollar prize for a proof of the conjecture, provided it was submitted before April 2002. The prize was never claimed, which mathematicians did not find surprising, because the problem has resisted all efforts for more than 250 years.

It began in 1742, when Christian Goldbach wrote to Leonhard Euler, suggesting that every even integer is the sum of two primes. (Apparently René Descartes had come across the same idea a little earlier, but no one had noticed.) At that time the number 1 was considered to be prime, so $2 = 1 + 1$ was acceptable, but nowadays we reformulate the the *Goldbach Conjecture* thus: every even integer greater than 2 is the sum of two primes – often in several different ways. For example,

$$4 = 2 + 2$$
$$6 = 3 + 3$$
$$8 = 5 + 3$$
$$10 = 7 + 3 = 5 + 5$$
$$12 = 7 + 5$$
$$14 = 11 + 3 = 7 + 7$$

Euler replied that he was sure Goldbach must be right, but he couldn't find a proof – and that remains true today. We do know that every even integer is the sum of at most six primes – proved by Olivier Ramaré in 1995. In 1973 Chen Jing-Run proved that every sufficiently large even integer is the sum of a prime and a semiprime (either a prime or a product of two primes).

In 1998 Jean-Marc Deshouillers, Yannick Saouter and Herman te Riele verified Goldbach's Conjecture for all even numbers up to 10^{14}. By 2007, Oliveira e Silva had improved that to 10^{18}, and his computations continue. If the Riemann Hypothesis (page 215) is true, then the Odd Goldbach Conjecture – that every odd integer greater than 5 is the sum of three primes – is a consequence of the 1998 result.

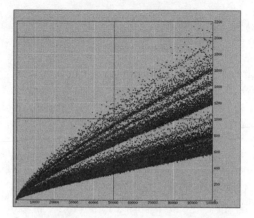

Graph showing in how many ways (vertical axis) a given even number (horizontal axis) can be expressed as a sum of two primes. The lowest points in the graph move upwards as we go from left to right, indicating that there are many ways to achieve this. However, for all we know an occasional point might fall on the horizontal axis. Just one such point would disprove the Goldbach Conjecture.

In 1923 Godfrey Hardy and John Littlewood obtained a heuristic formula – one that they could not prove rigorously, but looked plausible – for the number of different ways to write a given even integer as a sum of two primes. This formula, which agrees with numerical evidence, indicates that when the number gets large, there are many ways to write it as a sum of two primes. Therefore we may expect the smallest of the two primes to be relatively tiny. In 2001 Jörg Richstein observed that for numbers up to 10^{14}, the smaller prime is at most 5,569, and this occurs for

$$389,965,026,819,938 = 5,569 + 389,965,814,369$$

• •

Turtles All the Way Down

Infinity is a slippery idea. People talk fairly casually of 'eternity' – an infinite period of time. According to the Big Bang theory, the universe came into being about 13 billion years ago. Not only was there no universe before then – there was no 'before' before then.* Some people worry about that, and most of them seem much happier with the idea that the universe 'has always existed'. That is, its past has already been infinitely long.

This alternative seems to solve the difficult question of the origin of the universe, by denying that it ever had an origin. If something has always been here, it's silly to ask why it's here now. Isn't it?

Probably. But that still doesn't explain *why it's always been here*.

This can be a difficult point to grasp. To bring it into perspective, let me compare it with a rather different proposal. There is an amusing (and very likely true) tale that a famous scientist – Stephen Hawking is often mentioned because he told the story in *A Brief History of Time* – was giving a lecture about the universe, and a lady in the audience pointed out that the Earth floats in space because it rests on the back of four elephants, which in turn rest on the back of a turtle.

'Ah, but what supports the turtle?' the scientist asked.

'Don't be silly,' she said. '*It's turtles all the way down!*'

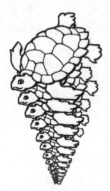

Turtles all the way down.

---

* Some cosmologists now think that there could have been something before the Big Bang after all – our universe may be part of a 'multiverse' in which individual universes could come into existence or fade away again. The theory is nice, but it's difficult to find any way to test it.

All very amusing, and we don't buy that explanation. A self-supporting pile of turtles is ludicrous, and not just because it's turtles. Each turtle being supported by a previous one just doesn't look like an explanation of how *the whole pile* stays up.

Very well. But now replace the Earth by the present state of the universe, and replace each turtle by the previous state of the universe. Oh, and change 'support' to 'cause'. Why does the universe exist? Because a previous one did. Why did that one exist? Because a previous one did. Did it all start a finite time in the past? No, it's *universes all the way back*.*

So a universe that has always existed is at least as puzzling as one that has not.

• •

Hilbert's Hotel

Among the paradoxes concerning the infinite are a series of bizarre events at Hilbert's Hotel. David Hilbert was one of the world's leading mathematicians around 1900. He worked in the logical foundations of mathematics and took a particular interest in infinity. Anyway, Hilbert's Hotel has infinitely many rooms, numbered 1, 2, 3, 4, and so on – every positive integer.

One bank holiday weekend, the hotel was completely full. A traveller without a reservation arrived at reception wanting a room. In any finite hotel, no matter how big, the traveller would be out of luck – but not in Hilbert's Hotel.

'No problem, sir,' said the manager. 'I'll ask the person in Room 1 to move to Room 2, the person in Room 2 to move to Room 3, the person in Room 3 to move to Room 4, and so on. The person in Room n will move to Room $n + 1$. Then Room 1 will be free, so I'll put you there.'

---

* Part of the appeal of the multiverse approach is that it revives the 'it's always been here' point of view. Our universe hasn't, but the surrounding multiverse has. It's multiverses all the way back …

1	2	3	4	5	6	7	8		n

1	2	3	4	5	6	7	8		n

All move up one, and Room 1 is free.

This trick works in an infinite hotel. In a finite hotel it goes wrong, because the person in the room with the biggest number has nowhere to go. But in Hilbert's Hotel there is *no* biggest room number. Problem sorted.

Ten minutes later, an Infinity Tours coach arrived, with infinitely many passengers sitting in seats 1, 2, 3, 4, and so on.

'Well, I can't fit you in by asking every other guest to move up some number of places,' said the manager. 'Even if they all moved up a million places, that would only free up a million rooms.' He thought for a moment. 'Nevertheless, I can still fit you in. I'll ask the person in Room 1 to move to Room 2, the person in Room 2 to move to Room 4, the person in Room 3 to move to Room 6, and so on. The person in Room n will move to Room $2n$. That frees up all the odd-numbered rooms, so now I can put the person in Seat 1 of your bus into Room 1, the person in Seat 2 into Room 3, the person in Seat 3 into Room 5, and so on. The person in Seat n will move to Room $2n - 1$.'

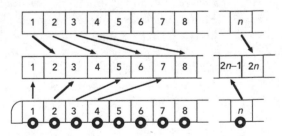

How to accommodate an infinite bus-load.

However, the manager's troubles were still not over. Ten minutes later, he was horrified to see infinitely many Transfinity Travel buses arriving in his (infinite) car park.

He rushed out to meet them. 'We're full – but I can *still* fit you all in!'

'How?' asked the driver of Bus 1.

'I'll reduce you to a problem I've already solved,' said the manager. 'I want you to move everyone into Bus 1.'

'But Bus 1 is full! And there are infinitely many other buses!'

'No problem. Line up all your buses side by side, and renumber all the seats using a diagonal order.'

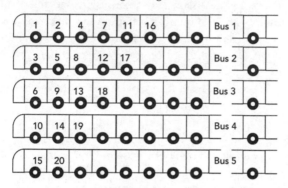

The Manager's 'diagonal' order – the numbers 2–3, 4–5–6, 7–8–9–10, and so on slant to the left.

'What does that achieve?' asked the driver.

'Nothing – yet. But notice: each passenger, in each of your infinitely many buses, is assigned a new number. Every number occurs exactly once.'

'And your point is—?'

'Move each passenger to the seat in Bus 1 that corresponds to their new number.'

The driver did so. Then everyone was sitting in Bus 1, and all the other buses were empty – so they drove away.

'Now I've got a full hotel and just one extra bus-load,' said the manager. 'And I already know how to deal with that.'

Continuum Coaches

You won't be surprised to hear that the Hilbert's Hotel eventually ran into an accommodation problem that the manager could *not* solve. This time the hotel was completely empty – not that this ever seemed to make much difference. Then one of Cantor's Continuum Coaches stopped at the front door.

Georg Cantor was the first to sort out the mathematics of infinite sets. And he discovered something remarkable about the 'continuum' – the real number system. A *real number* is one that can be written as a decimal, which can either stop after finitely many digits, like 1.44, or go on for ever, like π. Here's what Cantor found.

The seats of the Continuum Coach were numbered using real numbers, not positive integers.

'Well,' the manager thought, 'one infinity is just like any other, right?' So he assigned passengers to rooms, and eventually the Hotel was full and the lobby was empty. The manager sighed with relief. 'Everyone has a room,' he said to himself.

Then a forlorn figure came in through the revolving doors.

'Good evening,' said the manager.

'My name is Mr Diagonal. Geddit? Missed-a-diagonal. You've missed me out, mate.'

'Well, I can always bump everyone up one room—'

'No, mate, you said "Everyone has a room" – I heard you. But I don't.'

'Nonsense! You've gone to your room, then nipped out the back and come in the front. I know your kind!'

'No, mate – I can *prove* I'm not in any of your rooms. Who's in Room 1?'

'I can't reveal personal information about guests.'

'What's the first decimal place of their coach seat?'

'I suppose I can reveal that. It's a 2.'

'My first digit is 3. So I'm not the person in Room 1, mate. Agreed?'

'Agreed.'

'What's the *second* decimal place of the coach seat of the person in Room 2?'

'It's a 7.'

'My second digit is 5. So I'm not the person in Room 2.'

'That makes sense.'

'Yeah, mate, and it goes on doing that. What's the *third* decimal place of the coach seat of the person in Room 3?'

'It's a 4.'

'My third digit is 8. So I'm not the person in Room 3.'

'Hmm. I think I see where this is headed.'

'Too right, mate. My nth digit is different from the nth digit of the person in Room n, *for every* n. So I'm not in Room n. Like I said, you left me out.'

'And like *I* said, I can always bump everyone up one place and fit you in.'

'No use, mate. There's infinitely many more just like me out there, sitting in your car park waiting for a room. However you assign passengers to rooms, there's going to be someone on the coach whose nth digit is different from the nth digit of the person in Room n, for every n. Hordes of them, in fact. You'll always miss people out.'

Now, you understand that Cantor didn't quite write his proof in those terms, but that was the basic idea. He proved that the infinite set of real numbers can't be matched, one for one, with the infinite set of whole numbers. Some infinities are bigger than others.

• •

A Puzzling Dissection

'Why are you hacking that chessboard to bits?' asked Innumeratus.

'I want to show you something about areas,' said Mathophila. 'What's the area of the chessboard if each square has area one square unit?'

Innumeratus thought about this, and because he was better

at maths than his name might suggest, he quickly said, 'It's 8 times 8, which is 64 square units.'

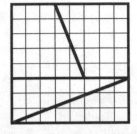

How Mathophila cut up
her chessboard . . .

'Excellent!' said Mathophila. 'Now, I'm going to rearrange the four pieces to make a rectangle.'

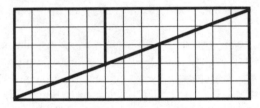

. . . and how she rearranged the pieces.

'OK,' said Innumeratus.

'What's the area of the rectangle?'

'Er – it must be 64 square units as well! It's made from the same pieces.'

'Right . . . but what size is the rectangle?'

'Let me see – 13 by 5.'

'And what is 13 times 5?'

'65,' replied Innumeratus. He paused. 'So its area must be 65 square units. That's strange. The area can't change when the pieces are reassembled in a different way . . .'

So what's happened?

Answer on page 286

A *Really* Puzzling Dissection

'The area can't change when the pieces are reassembled in a different way.'

Hmmm.

In 1924 two Polish mathematicians, Stefan Banach and Alfred Tarski, proved that it is possible to dissect a sphere into finitely many pieces, which can then be rearranged to make two spheres – each the same size as the original. No overlaps, no missing bits – the pieces fit together perfectly. This result has become known as the *Banach–Tarski paradox*, although it's a perfectly valid theorem and the only element of paradox is that it seems to be obviously false.

It *can* be done – but not with pieces like these.

Hang on, though. Surely, if you cut a sphere into several pieces, the total volume of the pieces must be the same as that of the sphere. So however you reassemble the pieces, the total volume can't change. But two identical spheres have twice the volume of a single sphere (of the same size). You don't have to be a genius to see that it can't be done! In fact, if it *could* be done, then you could start with a gold sphere, cut it up, fit the pieces back together, and end up with twice as much gold. Then repeat ... But you can't get something for nothing.

Hang on, though. Let's not be so hasty.

The argument about gold is inconclusive, because mathematical concepts do not always model the real world exactly. In mathematics, volumes can be subdivided into indefinitely small pieces. In the real world, you hit problems at the atomic scale. This could spoil things if we try to use gold.

In contrast, the argument about volumes looks watertight. But there's a tiny loophole in the logic: the tacit assumption that the separate pieces *have* well-defined volumes. 'Volume' is such a familiar concept that we tend to forget just how tricky it can be.

None of this means that Banach and Tarski were right; it just explains why they are not *obviously* wrong. Unlike the nice polygonal pieces in Mathophila's dissection of a chessboard, the Banach–Tarski 'pieces' are more like disconnected clouds of infinitely small dust specks than solid lumps. They are so complicated, in fact, that their volumes cannot be defined – not if we want them to obey the usual rule 'when you combine several pieces, their volumes add'. And if that rule fails, the argument about volumes comes to bits. The single sphere, and two copies of it, have well-defined volumes. But the intermediate stages, when they are cut into pieces, aren't like that.

What *are* they like? Well ... not like that.

Banach and Tarski realised that this loophole might actually make their paradoxical dissection possible. They proved that:

- You can split a single sphere A into finitely many very complicated, possibly disconnected, parts.
- You can do the same to two spheres B and C, the same size as A.
- You can accomplish all of that in such a way that the parts of B and C together correspond exactly to the parts of A.
- You can arrange for corresponding parts to be perfect copies of one another.

The proof of the Banach–Tarski paradox is complicated and technical, and it requires a set-theoretic assumption known as the axiom of choice. This particular assumption worries some mathematicians. However, the fact that it leads to the Banach–Tarski paradox is not what worries them, and is no reason to reject it. Why not? Because the Banach–Tarski paradox isn't really very paradoxical. With the right intuition, we would expect such paradoxical dissections to be possible.

Let me try to give you that intuition. It all hinges on the weird behaviour of what are called infinite sets. Although a sphere has finite size, it contains infinitely many *points*. That leaves room for the weirdness of infinity to show up in the geometry of the sphere.

A useful analogy involves the English alphabet, the 26 letters A, B, C, ..., Z. These letters can be combined to make *words*, and we list permissible words in a *dictionary*. Suppose we allow all possible sequences of letters, as long or as short as we like. So AAAAVDQX is a word, and so is GNU, and so is ZZZ...Z with ten million Z's. We can't print such a dictionary, but to mathematicians it is a well-defined set which contains infinitely many words.

Now, we can dissect this dictionary into 26 pieces. The first piece contains all words starting with A, the second contains all words starting with B, and so on, with the 26th piece containing all words starting with Z. These pieces do not overlap, and every word occurs in exactly one piece.

Each piece, however, has exactly the same structure as the original dictionary. The second piece, for example, contains the words BAAAAVDQX, BGNU and BZZZ...Z. The third contains CAAAAVDQX, CGNU and CZZZ...Z. You can convert each piece into the entire dictionary by lopping off the first letter from every word.

In other words: we can cut the dictionary apart, and reassemble the pieces to make 26 exact copies of the dictionary.

Banach and Tarski found a way to do the same kind of thing with the infinite set of all points in a solid sphere. Their alphabet consisted of two different rotations of the sphere; their words were sequences of these rotations. By playing a more complicated version of the dictionary game with the rotations, you can create an analogous dissection of the sphere. Since there are now two 'letters' in the alphabet, we convert the original sphere into two identical copies.

Careful readers will observe that I've cheated slightly in the interests of simplicity. When I chop off the initial letter B from the second piece, for instance, I not only get the entire original dictionary: I also get the 'empty' word that arises when the initial B is deleted from the word B. So really my dissection turns the dictionary into 26 copies of itself, plus 26 extra words A, B, C, ..., Z of length 1. To keep everything neat and tidy, we have to absorb the extra 26 words into the pieces. A similar problem occurs in Banach and Tarski's construction – but this is a very fine point. If we ignore it, we still double the sphere – we just have a few extra points *left over*. Which is just as suprising.

After Banach and Tarski proved their theorem, mathematicians began to wonder how few pieces you could get away with. In 1947 Abraham Robinson proved that it can be done with five pieces, but no fewer. If you are willing to ignore a single point at the centre of the sphere, this number reduces to four.

The Banach–Tarski paradox isn't really about dissecting spheres. It is about the impossibility of defining a sensible concept of 'volume' for really complicated shapes.

Nothing Up My Sleeve ...

How can you remove a loop of string from your arm without taking your hand out of your jacket pocket?

More precisely: take a two-metre length of string and tie its ends together to form a closed loop. Put on your jacket, button it up, and put your arm through a loop and into the side pocket. Now you have to remove the loop without taking your hand out of your pocket – and without sliding the loop into the pocket to sneak it out over the ends of your fingers.

Remove the string without removing your hand from your pocket.

Answer on page 286

Nothing Down My Leg ...

Once your audience has learned to solve the previous problem, ask someone to try the same thing, still wearing the jacket, but with his hand in his *trouser* pocket.

Answer on page 287

• •

Two Perpendiculars

Euclidean geometry is renowned for its logical consistency: no two theorems contradict each other. Actually, there are errors in Euclid. Here's a case in point.

One of Euclid's theorems proves that if we have a line, and a point not on the line, then there is *exactly one* 'perpendicular' from the point to the line. That is, there is a line through the point that meets the original line at right angles – and there is only one such line. (If there were two, they would be parallel, so they couldn't both pass through the same point.)

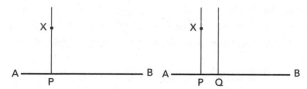

Given AB and X, we can find P such that PX is perpendicular to AB. There can't be another point Q like that, because the line through Q is parallel to PX so it can't pass through X.

A second Euclidean theorem proves that if you take a circle and join the two ends of a diameter to a point on the circumference, you get a right angle.

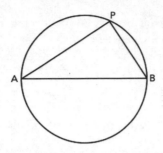

If AB is a diameter of the circle, angle APB is a right angle.

Let's put these two theorems together and see what happens.

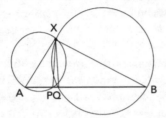

How to find two perpendiculars.

Given the line AB and the point X, draw circles with diameters AX and BX. Let the line AB meet the first circle at P and the second circle at Q. Then the angle APX is a right angle, since AX is a diameter of the first circle. Similarly, the angle BQX is a right angle. So there are *two* perpendiculars XP and XQ from X to AB.

Which of Euclid's two theorems is wrong?

Answer on page 288

● ●

Can You Hear the Shape of a Drum?

The backcloth depicted a striking scene: the Rhine valley by moonlight. In the pit, the orchestra was rehearsing Wagner's *Götterdämmerung*. The story had reached the tragic death of Siegfried, and the conductor, Otto Fenderbender, raised his baton for the beginning of the 'Funeral March'. First, just tympani, an intricate repeated rhythm in a low C sharp ...

'No, no, no!' screamed Fenderbender, hurling his baton to the floor. 'Not like that, you incompetent pigs!'

The leading tympanist, somewhat unwisely, protested. 'But Herr Fenderbender, the rhythm was absolutely pre—'

'Rhythm, schmythm!' said Fenderbender.

'The tempo was exactly as the score indic—'

'I am *not* complaining about the *tempo*!' screamed the conductor.

'The pitch was a perfect C shar—'

'Pitch? *Pitch?* Of *course* the pitch was perfect! I heard that for myself when the orchestra was tuning up! I have an inherent sense of pitch!'

'Then what—'

'The *shape*, you fool! The shape!'

The lead tympanist looked perplexed. It was hard to describe. Otto tried to express what he had heard. 'One of the drums sounded too ... well, too *square*,' he said. 'The other tympani had their usual ... *rounded* sound, but one of them— well, it had *corners*.'

'Come now, Herr Fenderbender – surely you're not claiming you can *hear* the shape of a drum?'

'I heard what I heard,' Otto said doggedly. 'One of the drums is too square.'

And, what do you know? He was right. It's the Bessel functions, you see.

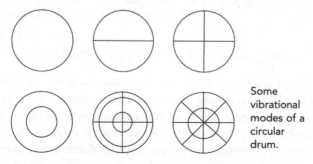

Some vibrational modes of a circular drum.

Let me explain. When a drum beats, it produces several

different notes at once, each note corresponding to a different *mode* of vibration. Each has its own frequency, or equivalently, pitch. Euler calculated the vibrational spectrum of a circular drum – the list of frequencies of these basic modes – using mathematical gadgets called Bessel functions. For a square drum you get sines and cosines instead. In both cases there are characteristic patterns of *nodal lines*, where the drum remains stationary. At any given instant, the drum is displaced upwards on one side of a nodal line, and downwards on the other. As the drum vibrates, each region between the nodal lines oscillates up and down. Fast oscillations create a high-pitched sound, slow oscillations make the pitch lower.

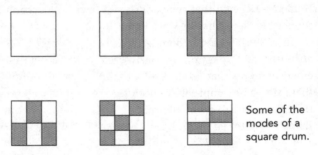

Some of the modes of a square drum.

The mathematics of vibrations proves that the shape of a drum determines its list of frequencies – basically, what it can sound like. But can we go the other way, and deduce the shape from the sound? In 1966 Mark Kac made that question precise: given the spectrum, is it possible to find the shape of the drum?

Kac's question is much more important than its quirky formulation might suggest. When an earthquake hits, the entire Earth rings like a bell, and seismologists deduce a great deal about the internal structure of our planet from the 'sound' that it produces and the way those sounds echo around as they bounce off different layers of rock. Kac's question is the simplest one we can ask about such techniques. 'Personally, I believe that one cannot "hear" the shape,' Kac wrote. 'But I may well be wrong and I am not prepared to bet large sums either way.'

The first significant evidence that Kac was right showed up in a higher-dimensional analogue of the problem. John Milnor wrote a one-page paper proving that two distinct 16-dimensional tori (generalised doughnuts, basically) have the same spectrum. The first results for ordinary 2-dimensional drums were in a more positive direction: various features of the shape *can* be deduced from the spectrum. Kac himself proved that the spectrum of a drum determines its area and perimeter. A curious consequence is that you can hear whether or not a drum is circular, because a circle has the smallest perimeter for a given area. If you know the area A and the perimeter p, and it so happens that $p^2 = 4\pi A$ – as it is for a circle – then the drum *is* a circle, and vice versa. So when Fenderbender said that tympani should have a nice 'rounded' sound, he knew what he was talking about.

In 1989 Carolyn Gordon, David Webb and Scott Wolpert answered Kac's question by constructing two distinct mathematical drums that produce an identical range of sounds. Since then, simpler examples have been found. So now we know that there are limits to what information can be deduced from a shape's vibrational spectrum.

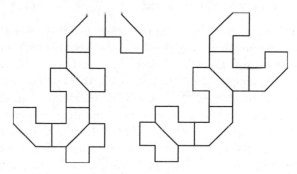

The first example of two sound-alike drums with different shapes.

What is e, and Why?

The number e, which is approximately 2.7182, is the 'base of natural logarithms', a term that refers to its historical origins. One way to see how it arises is to see how a sum of money grows when compound interest is applied at increasingly fine intervals. Suppose that you deposit £1 in the Bank of Logarithmania—

No, no, no. This is the twenty-first century. People don't deposit savings in banks, they borrow.

OK, suppose you borrow £1 on your Logarithmania credit card. (More likely it would be £4,675.23, but £1 is easier to think about.) Once the 0% balance transfer deal has lapsed – about a week after you sign up for the card – the bank applies an interest rate of 100%, paid annually. Then after one year you will owe them

$$£1.00 \text{ borrowed} + £1.00 \text{ interest} = £2.00 \text{ total}$$

If instead you paid 50% interest every six months, *compounded* (so that interest becomes payable on previous interest) then after one year you would owe

$$£1.00 \text{ invested} + £0.50 \text{ interest} + £0.75 \text{ interest} = £2.25 \text{ total}$$

This is $(1 + \frac{1}{2})^2$, and the pattern continues like that. So, for example, if you paid interest of 10% at intervals of one-tenth of a year, you would end up owing

$$\left(1 + \frac{1}{10}\right)^{10} = 2.5937$$

pounds. The Bank likes the way these sums are going, so it decides to apply the interest rate ever more frequently. If you paid interest of 1% at intervals of one-hundredth of a year, you would end up owing

$$\left(1 + \frac{1}{100}\right)^{100} = 2.7048$$

pounds. If you paid interest of 0.1% at intervals of one-thousandth of a year, you would end up owing

$$\left(1 + \frac{1}{1,000}\right)^{1,000} = 2.7169$$

pounds. And so on.

As the intervals become ever finer, the amount you owe does not increase without limit. It just seems that way. The amount owed gets closer and closer to 2.7182 pounds – and this number is given the symbol e. It's one of those weird numbers which, like π, turn up naturally in mathematics but can't be expressed exactly as a fraction, so it gets a special symbol. It is especially important in calculus, and it is widely used in scientific applications.

● ●

May Husband and Ay ...

In a single move, a chess queen can travel any number of squares in a straight line – horizontally, vertically or diagonally. (Unless another piece stops her, but we ignore that in this puzzle.)

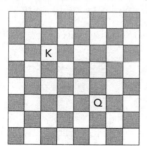

Move the queen from Q to K, visiting each square exactly once and in as few moves as possible.

She starts on square Q and wishes to visit the king on square K. Along the way, she wants to visit all of her other subjects, who live on the other 62 squares. Just passing through, you appreciate – she doesn't stop on every square, but she does has to stop now and again. How can she visit all the squares and finish at the

king's square, without passing through any square twice – in the
smallest number of moves?

Answer on page 288

• •

Many Knees, Many Seats

A polyhedron is a solid with finitely many flat (that is, planar)
faces. Faces meet along lines called *edges*; edges meet at points
called *vertices*. The climax of Euclid's *Elements* is a proof that there
are precisely five *regular polyhedrons*, meaning that every face is a
regular polygon (equal sides, equal angles), all faces are identical,
and each vertex is surrounded by exactly the same arrangement of
faces. The five regular polyhedrons (also called regular solids) are:

- the tetrahedron, with 4 triangular faces, 4 vertices and 6 edges
- the cube or hexahedron, with 6 square faces, 8 vertices and
  12 edges
- the octahedron, with 8 triangular faces, 6 vertices and 12 edges
- the dodecahedron, with 12 pentagonal faces, 20 vertices and
  30 edges
- the icosahedron, with 20 triangular faces, 12 vertices and
  30 edges.

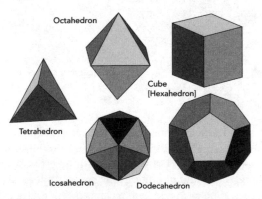

The five regular solids.

The names start with the Greek word for the number of faces, and 'hedron' means 'face'. Originally it meant 'seat', which isn't quite the same thing. While we're discussing linguistics, the '-gon' in 'polygon' originally meant 'knee' and later acquired the technical meaning of 'angle'. So a polygon has many knees, and a polyhedron has many seats.

The regular solids arise in nature – in particular, they all occur in tiny organisms known as radiolarians. The first three also occur in crystals; the dodecahedron and icosahedron don't, although irregular dodecahedra are sometimes found.

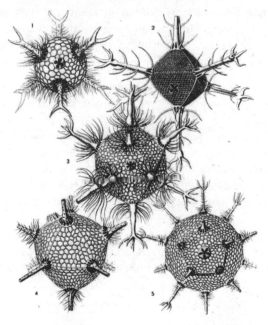

Radiolarians shaped like the regular solids.

It's quite easy to make models of polyhedrons out of card, by cutting out a connected set of faces – called the *net* of the solid – folding along edges, and gluing or taping appropriate pairs of edges together. It helps to add flaps to one edge of each such pair, as shown.

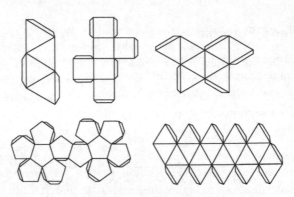

Nets of the regular solids.

Here's a bit of arcane lore: if the edges are of unit length, then the volumes of these solids (in cubic units) are:

- Tetrahedron: $\frac{\sqrt{2}}{12} \sim 0.117\,851$
- Cube: 1
- Octahedron: $\frac{\sqrt{2}}{3} \sim 0.471\,405$
- Dodecahedron: $\frac{\sqrt{5}}{2}\phi^4 \sim 7.663\,12$
- Icosahedron: $\frac{\sqrt{5}}{6}\phi^2 \sim 2.181\,69$

Here ϕ is the golden number (page 96), which turns up whenever you have pentagons around – just as π turns up whenever you have spheres or circles. And $\sim$ means 'approximately equals'.

Analogues of the regular polyhedrons can be defined in spaces of 4 or more dimensions, and are called *polytopes*. There are six regular polytopes in 4 dimensions, but only three regular polytopes in 5 dimensions or more.

• •

Euler's Formula

The regular solids have a curious pattern which turns out to be far more general. If F is the number of faces, E the number of edges and V the number of vertices, then

$$F - E + V = 2$$

for all five solids. In fact, the same formula holds for any polyhedron that has no 'holes' in it – one that is topologically equivalent to a sphere. This relation is called *Euler's formula*, and its generalisations to higher dimensions are important in topology.

The formula also applies to a map in the plane, provided we consider the infinite region outside the map to be an extra face – or ignore this 'face' and replace the formula by

$$F - E + V = 1$$

which amounts to the same thing but is easier to think about. I'll call this expression *Euler's formula for maps*.

The diagram shows, using a typical example, why this formula is true. The value of $F - E + V$ is written underneath each step in the process. The method of proof is to simplify the map, one step at a time. If we choose a face adjacent to the outside of the map, and remove that face and an adjacent outside edge, then both F and E decrease by 1. This leaves $F - E$ unchanged. Since we haven't altered V, it also leaves $F - E + V$ unchanged. We can keep erasing a face and a corresponding edge until all faces have been removed. We are left with a network of edges and vertices, and this always forms a 'tree' – there are no closed loops of edges. In the example in the diagram, this stage is reached at the sixth step, when $F - E + V = 0 - 7 + 8$.

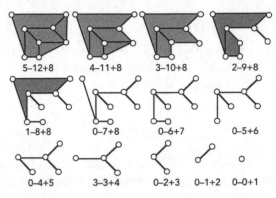

5–12+8   4–11+8   3–10+8   2–9+8

1–8+8   0–7+8   0–6+7   0–5+6

0–4+5   3–3+4   0–2+3   0–1+2   0–0+1

Proof of Euler's formula for maps in the plane.

Now we simplify the tree by snipping off one 'branch' – an edge on the end of the tree plus the vertex on the outside end of that edge – at a time. Now F remains at 0, while E and V both decrease by 1 with each snip. Again, $F - E + V$ remains unchanged. Eventually, a single vertex remains. Now $F = 0$, $E = 0$ and $V = 1$. So at the end of the process, $F - E + V = 1$. Since the process does not change this quantity, it must also have been equal to 1 when we started.

The proof explains why the signs alternate – plus, minus, plus – as we go from faces to edges to vertices. A similar trick works for higher-dimensional topology, for much the same reason.

There is a hidden topological assumption in the proof: the map is drawn in the plane. Equivalently, when considering polyhedrons, they must be 'drawable' on the surface of a sphere. If the polyhedron or map lives on a surface that is topologically distinct from a sphere, such as a torus, then the method of proof can be adapted but the final result is slightly different. For example, the formula for polyhedrons becomes

$$F - E + V = 0$$

when the polyhedron is topologically equivalent to a torus. As an example, this 'picture frame' polyhedron has $F = 16$, $E = 32$ and $V = 16$:

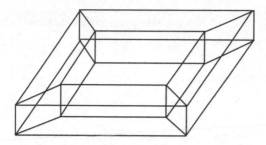

Picture frame polyhedron.

On a surface with g holes, the formula becomes

$$F - E + V = 2 - 2g$$

so we can calculate the number of holes by drawing a polyhedron on the surface. In this manner, an ant that inhabited the surface and could not perceive it 'from outside' would still be able to work out the topology of the surface. Today's cosmologists are trying to work out the topological shape of our own universe – which *we* can't observe 'from outside' – by using more elaborate topological ideas of a similar kind.

What Day is It?

Yesterday, Dad got confused about which day of the week it was. 'Whenever we go on holiday, I forget,' he said.

'Friday,' said Darren.

'Saturday,' his twin sister Delia contradicted.

'What day is it tomorrow, then?' asked Mum, trying to sort out the dispute without too much stress.

'Monday,' said Delia.

'Tuesday,' said Darren.

'Oh, for Heaven's sake! What day was it yesterday, then?'

'Wednesday,' said Darren.

'Thursday,' said Delia.

'Grrrrrrrr!' said Mum, doing her famous Marge Simpson

impression. 'Each of you has given one correct answer and two wrong ones.'

What day is it today?

Answer on page 288

. . . .

Strictly Logical

Only an elephant or a whale gives birth to a creature that weighs more than 100 kilograms.

The President weighs 150 kilograms.

Therefore ...

(I learned this one from the writer and publisher Stefan Themerson.)

. . . .

Logical or Not?

If pigs had wings, they'd fly.

Pigs don't fly if the weather is bad.

If pigs had wings, the sensible person would carry an umbrella.

Therefore:

If the weather is bad, the sensible person would carry an umbrella.

Is the deduction logically valid?

Answer on page 288

. . . .

A Question of Breeding

Farmer Hogswill went to the village fete, where he met five of his friends: Percy Catt, Dougal Dogge, Benjamin Hamster, Porky Pigge and Zoe Zebra. By a remarkable coincidence – which was a constant source of amusement – each of them was an expert breeder of one type of animal: cat, dog, hamster, pig and zebra. Between them, they bred all five types. None bred an animal that sounded like their surname.

'Congratulations, Percy!' said Hogswill. 'I hear you've just won third prize in the pig-breeding competition!'

'That's right,' said Zoe.

'And Benjamin got second for dogs!'

'No,' said Benjamin. 'You knows fine well I never touches no dogs. Nor zebras, neither.'

Hogswill turned to the person whose surname sounded like the animals that Zoe bred. 'And did you win anything?'

'Yes, a gold medal for my prize hamster.'

Assuming that all statements except the alleged second for dogs are true, who breeds what?

Answer on page 289

. . . .

Fair Shares

In 1944, as the Russian army fought to reclaim Poland from the Germans, the mathematician Hugo Steinhaus, trapped in the city of Lvov, sought distraction in a puzzle. As you do.

The puzzle was this. Several people want to share a cake (by all means replace that by a pizza if you wish). And they want the procedure to be fair, in the sense that no one will feel that they have got less than their fair share.

Steinhaus knew that for two people there is a simple method: one person cuts the cake into two pieces, and the other chooses which one they want. The second person can't complain, because they made the choice. The first person also can't complain – if they do, it was their fault for cutting the cake wrongly.

How can *three* people divide a cake fairly?

Answer on page 289

. . . .

The Sixth Deadly Sin

It's envy, and the problem is to avoid it.

Stefan Banach and Bronislaw Knaster extended Steinhaus's method of fair cake division to any number of people, and simplified it for three people. Their work pretty much summed up the whole area until a subtle flaw emerged: the procedure may be fair, but it takes no account of envy. A method is *envy-free* if no one thinks that anyone else has got a bigger share than they have. Every envy-free method is fair, but a fair method need not be envy-free. And neither Steinhaus's method, nor that of Banach and Knaster, is envy-free.

For example, Belinda may think that Arthur's division is fair. Then Steinhaus's method stops after step 3, and both Arthur and Belinda consider all three pieces to be of size 1/3. Charlie must think that his own piece is at least 1/3, so the allocation is proportional. But if Charlie sees Arthur's piece as 1/6 and Belinda's as 1/2, then he will envy Belinda, because Belinda got first crack at a piece that Charlie *thinks* is bigger than his.

Can you find an envy-free method for dividing a cake among three people?

Answer on page 290

Weird Arithmetic

'No, Henry, you can't do that,' said the teacher, pointing to Henry's exercise book, where he had written

$$\frac{1}{4} \times \frac{8}{5} = \frac{18}{45}$$

'Sorry, sir,' said Henry. 'What's wrong? I checked it on my calculator and it seemed to work.'

'Well, Henry, the *answer* is right, I guess,' the teacher admitted. 'Though you should probably cancel the 9's to get $\frac{2}{5}$, which is simpler. What's wrong is—'

Explain the mistake to Henry. Then find all such sums, with single non-zero digits in the first two fractions, that are correct.

Answer on page 291

. . . .

How Deep is the Well?

In one episode of the television series *Time Team*, the indefatigable archaeologists want to measure the depth of a mediaeval well. They drop something into it and time its fall, which takes an amazingly long six seconds. You hear it clattering its way down for ages. They come dangerously close to calculating the depth using Newton's laws of motion, but cop out at the last moment and use three very long tape measures joined together instead.

The formula they very nearly state is

$$s = \tfrac{1}{2}gt^2$$

where s is the distance travelled under gravity, falling from rest, and g is the acceleration due to gravity. It applies when air resistance can be ignored. This formula was discovered experimentally by Galileo Galilei and later generalised by Isaac Newton to describe motion under the influence of *any* force.

Taking $g = 10\,\mathrm{m\,s^{-2}}$ (metres per second per second), *how deep is the well?*

You've got three days to do it.

Answer on page 292

. . . .

McMahon's Squares

This puzzle was invented by the combinatorialist* P.A. McMahon in 1921. He was thinking about a square that has been divided into four triangular regions by diagonals. He wondered how many different ways there are to colour the various regions,

---

* A combinatorialist is someone who invents this kind of thing.

using three colours. He discovered that if rotations and reflections are regarded as the same colouring, there are exactly 24 possibilities. Find them all.

Now, a 6 × 4 rectangle contains 24 1 × 1 squares. Can you fit the 24 squares together to make such a rectangle, so that adjacent regions have the same colour, and the entire perimeter of the rectangle has the same colour?

Answer on page 292

● ●

What is the Square Root of Minus One?

The square root of a number is a number whose square is the given one. For instance, the square root of 4 is 2. If we allow negative numbers, then −2 is a second square root of 4 because minus times minus makes plus. Since plus times plus also makes plus, the square of any number – positive or negative – is always positive. So it looks as though negative numbers, in particular −1, can't have square roots.

Despite this, mathematicians (and physicists and engineers and indeed anyone working in any branch of science) have found it useful to provide −1 with a square root. This is not a number in the usual sense, so it is given a new symbol, which is i if you are a mathematician, and j if you are an engineer.

Square roots of negative numbers first showed up in mathematics around 1450, in an algebra problem. In those days the idea was a huge puzzle, because people thought of a number as something real. Even negative numbers caused a great deal of head-scratching, but people quickly got accustomed to them when they realised how useful they could be. Much the same happened with i, but it took a lot longer.

A big issue was how to visualise i geometrically. Everyone had got used to the idea of the number line, like an infinitely long ruler, with positive numbers on the right and negative ones on the left, and fractions and decimals in between:

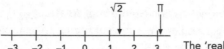

The 'real' number line.

Collectively, these familiar kinds of number became known as *real* numbers, because they correspond directly to physical quantities. You can *observe* 3 cows or 2.73 kilograms of sugar.

The puzzle was that there seemed to be nowhere on the real number line for the 'new' number i. Eventually, mathematicians realised that *it didn't have to go on the real number line*. In fact, being a new kind of number, it *couldn't* go there. Instead, i had to live on a second line, at right angles to the real number line:

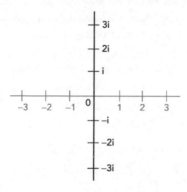

The 'imaginary' number line, placed at right angles to the real one.

And if you added an imaginary number to a real one, the answer had to live in the plane defined by the two lines:

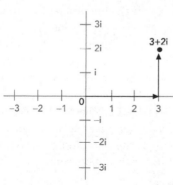

A *complex* number is a real one plus an imaginary one.

Multiplication was more complicated. The main point was that multiplying a number by i rotated it around the origin O through a right angle, anticlockwise. For instance, 3 multiplied by i is 3i, and that's what you get when you rotate the point labelled 3 through 90°.

The new numbers extended the familiar real number line to a larger space, a number plane. Three mathematicians discovered this idea independently: the Norwegian Caspar Wessel, the Frenchman Jean-Robert Argand and the German Carl Friedrich Gauss.

Complex numbers don't turn up in everyday situations, such as checking the supermarket bill or measuring someone for a suit. Their applications are in things like electrical engineering and aircraft design, which lead to technology that we can use without having to know the underlying mathematics.

The engineers and designers need to know it, though.

● ●

The Most Beautiful Formula

Occasionally people hold polls for the most beautiful mathematical formula of all time – yes, they really do, I'm not making this up, honest – and nearly always the winner is a famous formula discovered by Euler, which uses complex numbers to link the two famous constants e and π. The formula is

$$e^{i\pi} = -1$$

and it is extremely influential in a branch of maths known as complex analysis.

● ●

Why is Euler's Beautiful Formula True?

I often get asked whether there is a simple way to explain why Euler's formula $e^{i\pi} = -1$ is true. It turns out that there is, but some preparation is needed – about two years of undergraduate mathematics.

This is uncomfortably like the joke about the professor who says in a lecture that some fact is obvious, and when challenged goes away for half an hour and returns to say 'yes, it is obvious,' and then continues lecturing without further explanation. It just takes two years instead of half an hour. So I'm going to give you the explanation. Skip this bit if it doesn't make sense – but it does illustrate how higher mathematics sometimes gains new insights by putting different ideas together in unexpected ways. The necessary ingredients are some geometry, some differential equations and a bit of complex analysis.

The main idea is to solve the differential equation

$$\frac{\mathrm{d}z}{\mathrm{d}t} = \mathrm{i}z$$

where z is a complex function of time t, with the initial condition $z(0) = 1$. It is standard in differential equations courses that the solution is

$$z(t) = \mathrm{e}^{\mathrm{i}t}$$

Indeed, you can *define* the exponential function e^w this way.

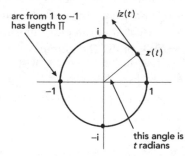

Geometry of the differential equation.

Now let's interpret the equation geometrically. Multiplication by i is the same as rotation through a right angle, so $\mathrm{i}z$ is at right angles to z. Therefore the tangent vector $\mathrm{i}z(t)$ to the solution at any point $z(t)$ is always at right angles to the 'radius vector' from 0 to $z(t)$ and has length 1. Therefore the solution $z(t)$ always lies in the unit circle, and the point $z(t)$ moves round this circle with angular velocity 1 measured in radians per second. (The *radian measure* of an angle is the length of the arc of the unit circle corresponding to that angle.) The circumference of the unit circle is 2π, so $t = \pi$ is halfway round the circle. But halfway round is visibly the point $z = -1$. Therefore $e^{i\pi} = -1$, which is Euler's formula.

All the ingredients of this proof are well known, but the overall package seems not to get much prominence. Its big advantage is to explain why circles (leading to π) have anything to do with exponentials (defined using e). So given the right background, Euler's formula ceases to be mysterious.

● ●

Your Call May be Monitored for Training Purposes

'The number you have dialled is imaginary. Please rotate your phone 90 degrees and try again.'

● ●

Archimedes, You Old Fraud!

'Give me a place to stand, and I will move the Earth.' So, famously, said Archimedes, dramatising his newly discovered law of the lever. Which in this case takes the form

> Force exerted by Archimedes
> × distance from Archimedes to fulcrum
> *equals*
> Mass of Earth × distance from Earth to fulcrum

The *fulcrum* is the pivot – the black triangle in the picture:

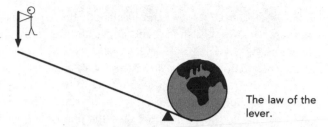

The law of the lever.

Now, I don't think Archimedes was interested in the position of the Earth in space, but he did want the fulcrum to be fixed. (I know he *said* 'a place to stand', but if the fulcrum moves, all bets are off, so presumably that's what he meant.) He also needed a perfectly rigid lever of zero mass, and he probably didn't realise that he also needed uniform gravity, contrary to astronomical fact, to convert mass to weight. No matter. I don't want to get into discussions about inertia or other quibbles. Let's grant him all those things. My question is: when the Earth moves, how *far* does it move? And can Archimedes achieve the same result more easily?

Answer on page 293

Fractals – The Geometry of Nature

Every so often, an entire new area of mathematics arises. One of the best known in recent times is *fractal geometry*, pioneered by Benoît Mandelbrot, who coined the term 'fractal' in 1975. Roughly speaking, it is a mathematical method for coming to grips with apparent irregularities in the natural world, and revealing hidden structure. The subject is best known for its beautiful, complex computer graphics, but it goes far deeper than that.

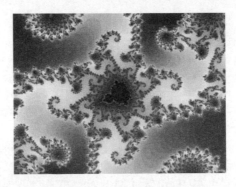

Part of the Mandelbrot set, a famous fractal.

The traditional shapes of Euclidean geometry are triangles, squares, circles, cones, spheres, and the like. These shapes are simple, and in particular they have no fine structure. If you magnify a circle, for instance, any portion of it looks more and more like a featureless straight line. Shapes like this have played a prominent role in science – for instance, the Earth is close to a sphere, and for many purposes that level of detail is good enough.

Many natural shapes are far more complex. Trees are a mass of branches, clouds are fuzzy and convoluted, mountains are jagged, coastlines are wiggly ... To understand these shapes mathematically, and to solve problems about them, we need new ingredients. The supply of problems, by the way, is endless – how do trees dissipate the energy of the wind, how do waves erode a coastline, how does water run off mountains into rivers? These are practical issues, often related to ecology and the environment, not just theoretical problems.

Coastlines are a good example. They are wiggly curves, but you can't use any old wiggly curve. Coastlines have a curious property: they look much the same on any scale of map. If the map shows more detail, extra wiggles can be distinguished. The exact shape changes, but the 'texture' seems pretty much the same. The jargon here is 'statistically self-similar'. All statistical features of a coastline, such as what proportion of bays have a given relative size, are the same no matter what scale of magnification you work on.

Mandelbrot introduced the word *fractal* to describe any shape that has intricate structure no matter how much you magnify it. It doesn't have to be statistically self-similar – but such fractals are easier to understand. And those that are exactly self-similar are even nicer, which is how the subject started.

About a century ago, mathematicians invented a spate of weird shapes, for various esoteric purposes. These shapes were not just statistically self-similar – they were exactly self-similar. When suitably magnified, the result looked identical to the original. The most famous is the *snowflake curve*, invented by Helge von Koch in 1904. It can be assembled from three copies of the curve shown in the right-hand diagram.

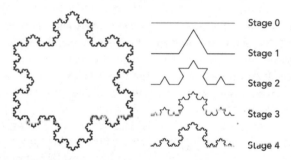

The snowflake curve and successive stages in its construction.

This component curve (though not the whole snowflake) is exactly self-similar. You can see that each stage in the construction is made from four copies of the previous stage, each one-third as big. The four copies are fitted together as in Stage 1. Passing to the infinite limit, we obtain an infinitely intricate curve that is built from four copies of itself, each one-third the size – so it is self-similar.

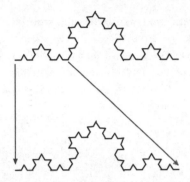

Each quarter of the curve, blown up to three times the size, looks like the original curve.

This shape is too regular to represent a real coastline, but it has about the right degree of wiggliness, and less regular curves formed in a similar way do look like genuine coastlines. The degree of wiggliness can be represented by a number, called the *fractal dimension*.

To see how this goes, I'm going to take some simpler *non-fractal* shapes and see how they fit together at different scales of magnification. If I break a line into pieces 1/5 the size, say, then I need 5 of them to reconstruct the line. With a square, I need 25 pieces, which is 5^2. And with cubes I need 125, which is 5^3.

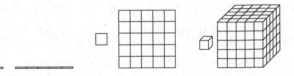

Effect of scaling on 'cubes' in 1, 2 and 3 dimensions.

The power of 5 occurring here is the same as the dimension of the shape concerned: 1 for a line, 2 for a square, 3 for a cube. In general, if the dimension is d and we have to fit k pieces of size $1/n$ together to reassemble the original shape, then $k = n^d$. Taking logarithms, we find that

$$d = \frac{\log k}{\log n}$$

Now, taking a deep breath, we try this formula out on the snowflake. Here we need $k = 4$ pieces each 1/3 the size, so $n = 3$. Therefore our formula yields

$$d = \frac{\log 4}{\log 3}$$

which is roughly 1.2618. So the 'dimension' of the snowflake curve is not a whole number!

That would be bad if we wanted to think of 'dimension' in the conventional way, as the number of independent directions available. But it's fine if we want a numerical measure of wiggliness, based on self-similarity. A curve with dimension 1.2618 is more wiggly than a curve of dimension 1, such as a straight line; but it is less wiggly than a curve of dimension 1.5, say.

There are dozens of technically distinct ways to define the dimension of a fractal. Most of them work when it is not self-similar. The one used by mathematicians is called the *Hausdorff–Besicovitch dimension*. It's a pig to define and a pig to calculate, but it has pleasant properties. Physicists generally use a simpler version, called the *box dimension*. This is easy to calculate, but lacks nearly all the pleasant properties of the Hausdorff–Besicovitch dimension. Despite that, the two dimensions are often the same. So the term *fractal dimension* is used to mean either of them.

Fractals need not be curves: they can be highly intricate surfaces or solids, or higher-dimensional shapes. The fractal dimension then measures how *rough* the fractal is, and how effectively it fills space. The fractal dimension turns up in most applications of fractals, both in the theoretical calculations and in experimental tests. For example, the fractal dimension of real coastlines is generally close to 1.25 – surprisingly close to that of the snowflake curve.

Fractals have come a long way, and they are now routinely used as mathematical models throughout the sciences. They are also the basis of an effective method for compressing computer files of video images. But their most interesting role is as 'the' geometry of many natural forms. A striking example is a kind of cauliflower called Romanesco broccoli. You can find it in most supermarkets. Each tiny floret has much the same form as the whole cauliflower, and everything is arranged in a series of ever-smaller Fibonacci spirals. This example is the tip of an iceberg – the fractal structure of plants. While much remains to be sorted out, it is already clear that the fractal structure arises from the way plants grow, which in turn is regulated by their genetics. So the geometry here is more than just a visual pun.

Romanesco broccoli – you can't get much more self-similar than that!

The applications of fractals are extensive, ranging from the fine structure of minerals to the form of the entire universe. Fractal shapes have been used to make antennas for mobile phones – such shapes are more efficient. Fractal image compression techniques cram huge quantities of data on to CDs and DVDs. There are even medical applications: for example, fractal geometry can be used to detect cancerous cells, whose surfaces are wrinkly and have a higher fractal dimension than normal cells.

About ten years ago a team of biologists (Geoffrey West, James Brown and Brain Enquist) discovered that fractal geometry can explain a long-standing puzzle about patterns in living creatures. The patterns concerned are statistical 'scaling laws'. For example, the metabolic rates of many animals seem to be proportional to the $\frac{3}{4}$th power of their masses, and the time it takes for the embryo to develop is proportional to the $-\frac{1}{4}$th power of the mass of the adult. The main enigma here is that fraction $\frac{1}{4}$. A power law with the value $\frac{1}{3}$ could be explained in terms of volume, which is proportional to the cube of the creature's length. But $\frac{1}{4}$, and related fractions such as $\frac{3}{4}$ or $-\frac{1}{4}$, are harder to explain.

The team's idea was an elegant one: a basic constraint on how organisms can grow is the transport of fluids, such as blood, around the body. Nature solves this problem by building a branching network of veins and arteries. Such a network obeys three basic rules: it should reach all regions of the body, it should transport fluids using as little energy as possible, and its smallest tubes should all be much the same size (because the tube can't be smaller than a single blood cell, or the blood can't flow). What shapes satisfy these conditions? Space-filling fractals – with the fine structure cut off at the limiting size, that of the single cell. This approach – which takes into account some important physical and biological details, such as the flexibility of the tubes and the occurrence of pulses in the blood as the heart beats – predicts that elusive $\frac{1}{4}$th power.

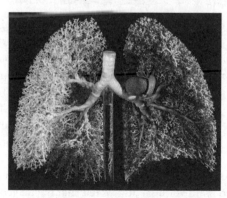

Fractal branching of blood vessels in the lungs.

196 // The Missing Symbol

The Missing Symbol

Place a standard mathematical symbol between 4 and 5 to get a number greater than 4 and less than 5.

Answer on page 294

. . . .

Where There's a Wall, There's a Way

In the county of Hexshire, fields are separated by walls built from the local stones – which for some reason are all made from identical hexagonal lumps joined together. Perhaps they originated as basalt columns like the ones in the Giant's Causeway. Anyway, Farmer Hogswill has seven stones, each formed from four hexagons. In fact, they are precisely the seven possible combinations of four hexagons:

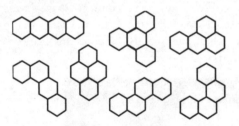

Seven stones to make a wall.

He has to make a wall shaped like this:

The required wall.

How can he do it? (He can rotate the stones and turn them over to obtain their mirror images if necessary.)

Answer on page 294

. . . .

Constants to 50 Places

π   3.141 592 653 589 793 238 462 643 383 279 502 884
    197 169 399 375 11
e   2.718 281 828 459 045 235 360 287 471 352 662 497
    757 247 093 699 96
√2 1.414 213 562 373 095 048 801 688 724 209 698 078
    569 671 875 376 95
√3 1.732 050 807 568 877 293 527 446 341 505 872 366
    942 805 253 810 38
log 2 0.693 147 180 559 945 309 417 232 121 458 176 568
    075 500 134 360 26
φ   1.618 033 988 749 894 848 204 586 834 365 638 117
    720 309 179 805 76
γ   0.577 215 664 901 532 860 606 512 090 082 402 431
    042 159 335 939 94
δ   4.669 201 609 102 990 671 853 203 820 466 201 617
    258 185 577 475 76

Here φ is the golden number (page 96), γ is Euler's constant (page 96), and δ is the *Feigenbaum constant*, which is important in chaos theory (page 117). See
en.wikipedia.org/wiki/Logistic_map
mathworld.wolfram.com/FeigenbaumConstant.html

• •

Richard's Paradox

In 1905 Jules Richard, a French logician, invented a very curious paradox. In the English language, some sentences define positive integers and others do not. For example 'The year of the Declaration of Independence' defines 1776, whereas 'The historical significance of the Declaration of Independence' does not define a number. So what about this sentence: 'The smallest number that cannot be defined by a sentence in the English language containing fewer than 20 words.' Observe that whatever this number may be, we have just defined it using a sentence in the English language containing only 19 words. Oops.

A plausible way out is to say that the proposed sentence does not actually define a specific number. However, it ought to. The English language contains a finite number of words, so the number of sentences with fewer than 20 words is itself finite. Of course, many of these sentences make no sense, and many of those that do make sense don't define a positive integer – but that just means that we have fewer sentences to consider. Between them, they define a finite set of positive integers, and it is a standard theorem of mathematics that in such circumstances there is a unique smallest positive integer that is not in the set. So on the face of it, the sentence does define a specific positive integer.

But logically, it can't.

Possible ambiguities of definition such as 'A number which when multiplied by zero gives zero' don't let us off the logical hook. If a sentence is ambiguous, then we rule it out, because an ambiguous sentence doesn't *define* anything. Is the troublesome sentence ambiguous, then? Uniqueness is not the issue: there can't be *two* distinct smallest-numbers-not-definable-(etc.), because one must be smaller than the other.

One possible escape route involves how we decide which sentences do or do not define a positive integer. For instance, if we go through them in some kind of order, excluding bad ones in turn, then the sentences that survive depend on the order in which they are considered. Suppose that two consecutive sentences are:

(1) The number in the next valid sentence plus one.

(2) The number in the previous valid sentence plus two.

These sentences cannot both be valid – they would then contradict each other. But once we have excluded one of them, the other one *is* valid, because it now refers to a different sentence altogether.

Forbidding this type of sentence puts us on a slippery slope, with more and more sentences being excluded for various reasons. All of which strongly suggests that the alleged sentence does not, in fact, define a specific number – even though it seems to.

• •

Connecting Utilities

Three houses have to be connected to three utility companies – water, gas and electricity. Each house must be connected to *all three* utilities. Can you do this without the connections crossing?

(Work 'in the plane' – there is no third dimension in which pipes can be passed over or under cables. And you are not allowed to route cables or pipes through a house or a utility company building.)

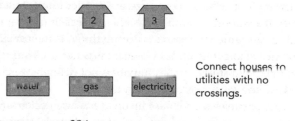

Connect houses to utilities with no crossings.

Answer on page 294

• •

Are Hard Problems Easy?
or
How to Win a Million Dollars by Proving the Obvious

Naturally, it's not *that* obvious. TANSTAAFL, as science fiction author Robert A. Heinlein used to say – There Ain't No Such Thing As A Free Lunch. But we can all dream.

I'm referring here to one of the seven Millennium Prize

Problems (page 127), whose solution will leave some lucky person a million dollars better off. Technically, it is known as 'P=NP?' which is a pretty silly name. But what it's about is of vital importance: inherent limits to the efficiency of computers.

Computers solve problems by running programs, which are lists of instructions. A program that always stops with the right answer (assuming that the computer is doing what its designers think it should) is called an 'algorithm'. The name honours the Arabic mathematician Abu Ja'far Muhammad ibn Musa al-Khwarizmi, who lived around AD 800 in present-day Iraq. His book *Hisab al-jabr w'al-muqabala* gave us the word 'algebra', and it consists of a series of procedures – algorithms – for solving algebraic equations of various kinds.

An algorithm is a method for solving a specific type of problem, but it is useless in practice unless it delivers the answer reasonably quickly. The theoretical issue here is not how fast the computer is, but how many calculations the algorithm has to perform. Even for a specific problem – to find the shortest route that visits a number of cities in turn, say – the number of calculations depends on how complicated the question is. If there are more cities to visit, the computer will have to do more work to find an answer.

For these reasons, a good way to measure the efficiency of an algorithm is to work out how many computational steps it takes to solve a problem of a given size. There is a natural division into 'easy' calculations, where the size of the calculation is some fixed power of the input data, and 'hard' ones, where the growth rate is much faster, often exponential. Multiplying two n-digit numbers together, for example, can be done in about n^2 steps using good old-fashioned long multiplication, so this calculation is 'easy'. Finding the prime factors of an n-digit number, on the other hand, takes about 3^n steps if you try every possible divisor up to the square root of n, which is the most obvious approach, so this calculation is 'hard'. The algorithms concerned are said to run in *polynomial time* (class P) and *non-polynomial time* (not-P), respectively.

Working out how quickly a given algorithm runs is relatively straightforward. The hard bit is to decide whether some other algorithm might be faster. The hardest of all is to show that what you've got is the fastest algorithm that will work, and basically we don't know how to do that. So problems that we think are hard might turn out to be easy if we found a better method for solving them, and this is where the million dollars comes in. It will go to whoever manages to prove that some specific problem is unavoidably hard – that no polynomial-time algorithm exists to solve it. Or, just possibly, to whoever proves that There Ain't No Such Thing As A Hard Problem – though that doesn't seem likely, the universe being what it is.

Before you rush out to get started, though, there are a couple of things you should bear in mind. The first is that there is a 'trivial' type of problem that is automatically hard, simply because the size of the *output* is gigantic. 'List all ways to rearrange the first n numbers' is a good example. However fast the algorithm might be, it takes at least $n!$ steps to print out the answer. So this kind of problem has to be removed from consideration, and this is done using the concept of a *non-deterministic polynomial time*, or NP, problem. (Note that NP is different from not-P.) These are the problems where you can verify a proposed *answer* in polynomial time – that is, easily.

My favourite example of an NP problem is solving a jigsaw puzzle. It may be very hard to find a solution, but if someone shows you an allegedly completed puzzle you can tell instantly whether they've done it right. A more mathematical example is finding a factor of a number: it is much easier to divide out and see whether some number works than it is to find that number in the first place.

The P=NP? problem asks whether every NP problem is P. That is, if you can check a proposed answer easily, can you *find* it easily? Experience suggests very strongly that the answer should be 'no' – the hard part is to find the answer. But, amazingly, no one knows how to prove that, or even whether it's correct. And that's why you can pocket a million bucks for proving that P is different from NP, or indeed for proving that, on the contrary, the two are equal.

As a final twist, it turns out that all likely candidates to show that P ≠ NP are in some sense equivalent. A problem is called *NP-complete* if a polynomial-time algorithm to solve that particular problem automatically leads to a polynomial-time algorithm to solve *any* NP problem. Almost any reasonable candidate for proving that P ≠ NP is known to be NP-complete. The nasty consequence of this fact is that no particular candidate is likely to be more approachable than any of the others – they all live or die together. In short: we know *why* P=NP? must be a very hard problem, but that doesn't help us to solve it.

I suspect that there are far easier ways to make a million.

• •

Don't Get the Goat

There used to be an American game show, hosted by Monty Hall, in which the guest had to choose one of three doors. Behind one was an expensive prize – a sports car, say. Behind the other two were booby prizes – goats.

After the contestant had chosen, Hall would open one of the *other* doors to reveal a goat. (With two doors to choose from, he could always do this – he knew where the car was.) He would then offer the contestant the chance to change their mind and choose the other unopened door.

Hardly anyone took this opportunity – perhaps with good reason, as I'll eventually explain. But for the moment let's take the problem at face value, and assume that the car has equal probability (one in three) of being behind any given door. We'll assume also that everyone knows ahead of time that Hall *always* offers the contestant a chance to change their mind, after revealing a goat. Should they change?

The argument against goes like this: the two remaining doors are equally likely to conceal a car or a goat. Since the odds are fifty–fifty, there's no reason to change.

Or is there?

Answer on page 296

• •

All Triangles are Isosceles

This puzzle requires some knowledge of Euclidean geometry, which nowadays isn't taught … Ho hum. It's still accessible if you're prepared to take a few facts on trust.

An *isosceles* triangle has two sides equal. (The third could also be equal: this makes the triangle *equilateral*, but it still counts as isosceles too.) Since it is easy to draw triangles with all three sides different, the title of this section is clearly *false*. Nevertheless, here is a geometric proof that it is true.

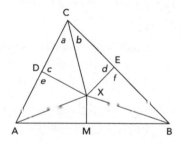

This triangle is isosceles – except that it clearly isn't.

(1) Take any triangle ABC.

(2) Draw a line CX that cuts the top angle in half, so that angles *a* and *b* are equal. Draw a line MX at right angles to the bottom edge at its midpoint, so that AM = MB. This meets the previous line, CX, somewhere inside the triangle at the point X.

(3) Draw lines from X to the other two corners A and B. Draw XD and XE to make angles *c*, *d*, *e* and *f* all right angles.

(4) Triangles CXD and CXE are *congruent* – that is, they have the same shape and size (though one is the other flipped over). The

reason is that angles *a* and *b* are equal, angles *c* and *d* are equal, and the side CX is common to both triangles.

(5)   Therefore lines CD and CE are equal.

(6)   So are lines XD and XE.

(7)   Since M is the midpoint of AB and MX is at right angles to AB, the lines XA and XB are equal.

(8)   But now triangles XDA and XEB are congruent. The reason is that XD = XE, XA = XB and angle *e* equals angle *f*.

(9)   Therefore DA = EB.

(10)   Combining steps 5 and 9: CA = CD + DA = CE + EB = CB. So lines CA and CB are equal, and triangle ABC is isosceles.

What's wrong here? (Hint: it's not the use of congruent triangles.)

   Answer on page 298

• •

Square Year

It was midnight on 31 December 2001, and Alfie and Betty – both of whom were aged less than sixty – were talking about the calendar.

   'At some time in the past, the year was the square of my father's age,' said Betty proudly. 'He died at the age of a hundred!'

   'And at some time in the future, the year will be the square of *my* age,' Alfie replied. 'I don't know whether I'll reach a hundred, though.'

   In which years were Betty's father and Alfie born?

   Answer on page 298

• •

Gödel's Theorems

In 1931 the mathematical logician Kurt Gödel proved two important theorems, of great originality, which placed unavoidable limits on the power of formal reasoning in mathematics. Gödel was responding to a research programme initiated by David Hilbert, who was convinced that the whole of mathematics could be placed on an axiomatic basis. Which is to say it should be possible to state a list of basic assumptions, or 'axioms', and deduce the rest of mathematics from the axioms. Additionally, Hilbert expected to be able to prove two key properties:

- The system is *logically consistent* – it is not possible to deduce two statements that contradict each other.
- The system is *complete* – every statement has either a proof or a disproof.

The kind of axiomatic 'system' that Hilbert had in mind was more basic than, say, arithmetic – something like the theory of sets introduced by Georg Cantor in 1879 and developed over the next few years. Starting from sets, there are ways to define whole numbers, the usual operations of arithmetic, negative and rational numbers, real numbers, complex numbers, and so on. So placing set theory on an axiomatic basis would automatically do the same for the rest of mathematics. And proving that the axiomatic system for set theory is consistent and complete would also do the same for the rest of mathematics. Since set theory is conceptually simpler than arithmetic, this seemed a sensible way to proceed. In fact, there was even a candidate axiomatisation of set theory, developed by Betrand Russell and Alfred North Whitehead in their three-volume epic *Principia Mathematica*. There were various alternatives, too.

Hilbert pushed a substantial part of his programme through successfully, but there were still some gaps when Gödel arrived on the scene. Gödel's 1931 paper 'On Formally Undecidable Propositions in *Principia Mathematica* and Related Systems I' left

Hilbert's programme in ruins, by proving that no such approach could ever succeed.

Gödel went to great lengths to place his proofs in a rigorous logical context, and to avoid several subtle logical traps. In fact most of his paper is devoted to setting up these background ideas, which are very technical – 'recursively enumerable sets'. The climax to the paper can be stated informally, as two dramatic theorems:

- In a formal system that is rich enough to include arithmetic, there exist *undecidable* statements – statements that can neither be proved nor disproved within that system.
- If a formal system that is rich enough to include arithmetic is logically consistent, then it is impossible to prove its consistency within that system.

The first theorem does not just indicate that finding a proof or disproof of the appropriate statement is difficult. It established that no proof exists, *and* no disproof exists. It means that the logical distinction between 'true' and 'false' is *not* identical to that between 'provable' and 'disprovable'. In conventional logic – including that used in *Principia Mathematica* – every statement is either true or false, and cannot be both. Since the negation not-P of any true statement P is false, and the negation of a false statement is true, conventional logic obeys the 'law of the excluded middle': given any statement P, then exactly one of P and not-P is true, and the other is false. Either $2+2$ is equal to 4, or $2+2$ is not equal to 4. It has to be one or the other, and it can't be both.

Now, if P has a proof, then P must be true – this is how mathematicians establish the truth (in a mathematical sense) of their theorems. If P has a disproof, then not-P must be true, so P must be false. But Gödel proved that for some statements P, neither P nor not-P has a proof. So a statement can be provable, disprovable – or *neither*. If it's neither, it is said to be 'undecidable'. So now there *is* a third possibility, and the 'middle' is no longer excluded.

Before Gödel, mathematicians had happily assumed that anything true was provable, and anything false was disprovable. *Finding* the proof or disproof might be very hard, but there was no reason to doubt that one or the other must exist. So mathematicians considered 'provable' to be the same as 'true', and 'disprovable' to be the same as 'false'. And they felt happier with practical concepts of proof and disproof than with deep and tricky philosophical concepts like truth and falsity, so mostly they settled for proofs and disproofs. And so it was disturbing to discover that these left a gap, a kind of logical no-man's-land. And in ordinary arithmetic, too!

Gödel set up his undecidable statement by finding a formal version of the logical paradox 'this statement is false', or more accurately of 'this statement has no proof'. However, in mathematical logic a statement is not permitted to refer to itself – in fact, 'this statement' is not something that has a meaning within the formal system concerned. Gödel found a cunning way to achieve much the same result without breaking the rules, by associating a numerical *code* with each formal statement. Then a proof of any statement corresponded to some sequence of transformations of the corresponding code number. So the formal system could model arithmetic – but arithmetic could also model the formal system.

Within this set-up, and assuming the formal system to be logically consistent, the statement P whose interpretation was basically 'this statement has no proof' must be undecidable. If P has a proof, then P is true, so by its defining property P has no proof – a contradiction. But the system is assumed to be consistent, so that can't happen. On the other hand, if P has no proof, then P is true. Therefore not-P has no proof. So neither P nor not-P has a proof.

From here it is a short step to the second theorem – if the formal system is consistent, then there can't be a proof that it is. I've always thought this to be rather plausible. Think of arithmetic as a used car salesman. Hilbert wanted to ask the salesman 'are you honest?' and get an answer that guaranteed that he was. Gödel basically argued that if you ask him this question and he says 'Yes, I am,' that is no guarantee of honesty. Would *you* believe that someone is telling the truth because they tell you they are? A court of law certainly would not.

Because of the technical complications, Gödel proved his theorems within one specific formal system for arithmetic, the one in *Principia Mathematica*. So a possible consequence might have been that this system is inadequate, and something better is needed. But Gödel pointed out in the introduction to his paper that a similar line of reasoning would apply to any alternative formal system for arithmetic. Changing the axioms wouldn't help. His successors filled in the necessary details, and Hilbert's programme was a dead duck.

Several important mathematical problems are now known to be undecidable. The most famous is probably the halting problem for Turing machines – which in effect asks for a method to determine in advance whether a computer program will eventually stop with an answer, or go on for ever. Alan Turing proved that some programs are undecidable – there is no way to prove that they stop, and no way to prove that they don't.

. . . .

If π isn't a Fraction, How Can You Calculate It?

The school value 22/7 for π is not exact. It's not even terribly good. But it *is* good for something so simple. Since we know that π is not an exact fraction, it's not obvious how it can be calculated to very high accuracy. Mathematicians achieve this using a variety of cunning formulas for π, all of which are exact, and all of which involve some process that goes on for ever. By stopping before we *get* to 'for ever', a good approximation to π can be found.

In fact, mathematics presents us with an embarrassment of riches, because one of the perennial fascinations of π is its tendency to appear in a huge variety of beautiful formulas. Typically they are infinite series, infinite products or infinite fractions (indicated by the dots ...) – which should not be a surprise since there is no simple finite expression for π, unless you cheat with integral calculus. Here are a few of the high points.

The first formula was one of the earliest expressions for π, discovered by François Viète in 1593. It is related to polygons with $2n$ sides:

$$\frac{2}{\pi} = \sqrt{\frac{1}{2}} \times \sqrt{\frac{1}{2} + \frac{1}{2}\sqrt{\frac{1}{2}}} \times \sqrt{\frac{1}{2} + \frac{1}{2}\sqrt{\frac{1}{2} + \frac{1}{2}\sqrt{\frac{1}{2}}}} \times \cdots$$

The next was found by John Wallis in 1655:

$$\frac{\pi}{2} = \frac{2}{1} \times \frac{2}{3} \times \frac{4}{3} \times \frac{4}{5} \times \frac{6}{5} \times \frac{6}{7} \times \frac{8}{7} \times \frac{8}{9} \times \cdots$$

Around 1675, James Gregory and Gottfried Leibniz both discovered

$$\frac{\pi}{4} = 1 - \frac{1}{3} + \frac{1}{5} - \frac{1}{7} + \frac{1}{9} - \frac{1}{11} + \frac{1}{13} - \cdots$$

This converges too slowly to be of any help in calculating π; that is, a good approximation requires oodles of terms. But closely related series were used to find several hundred digits of π in the eighteenth and nineteenth centuries. In the seventeenth century, Lord Brouncker discovered an infinite 'continued fraction':

$$\pi = \cfrac{4}{1 + \cfrac{1^2}{2 + \cfrac{3^2}{2 + \cfrac{5^2}{2 + \cfrac{7^2}{2 + \cdots}}}}}$$

and Euler discovered a pile of formulas like these:

$$\pi^2 = 1 + \frac{1}{2^2} + \frac{1}{3^2} + \frac{1}{4^2} + \frac{1}{5^2} + \frac{1}{6^2} + \cdots$$

$$\frac{\pi^3}{32} = 1 - \frac{1}{3^3} + \frac{1}{3^3} - \frac{1}{7^3} + \frac{1}{9^3} - \frac{1}{11^3} + \cdots$$

$$\frac{\pi^4}{90} = 1 + \frac{1}{2^4} + \frac{1}{3^4} + \frac{1}{4^4} + \frac{1}{5^4} + \frac{1}{6^4} + \cdots$$

(By the way, there seems to be no such formula for

$$1 + \frac{1}{2^3} + \frac{1}{3^3} + \frac{1}{4^3} + \frac{1}{5^3} + \frac{1}{6^3} + \cdots$$

which is very mysterious and not fully understood. In particular, this sum is not any simple rational number times π^3. We do know that the sum of the series is irrational.)

For the other formulas, we'll need the 'sigma notation' for sums. The idea is that: we can write the series for $\pi^2/6$ in the more compact form

$$\frac{\pi^2}{6} = \sum_{n=1}^{\infty} \frac{1}{n^2}$$

Let me unpack this. The fancy Σ symbol is Greek capital sigma, for 'sum', and it tells you to add together all the numbers to its right, namely $1/n^2$. The '$n=1$' below the Σ says that we start adding from $n=1$, and by convention n runs through the positive integers. The symbol ∞ over the Σ which means 'infinity', tells us to keep adding these numbers for ever. So this is the same series for $\pi^2/6$ that we saw earlier, but written as an instruction 'Add the terms $1/n^2$, for $n = 1, 2, 3$, and so on, going on for ever.'

Around 1985, Jonathan and Peter Borwein discovered the series

$$\frac{1}{\pi} = \frac{2\sqrt{2}}{9,801} \sum_{n=0}^{\infty} \frac{(4n)!}{(n!)^4} \times \frac{1,103 + 26,390n}{(4 \times 99)^{4n}}$$

which converges extremely rapidly. In 1997 David Bailey, Peter Borwein and Simon Plouffe found an unprecedented formula,

$$\pi = \sum_{n=0}^{\infty} \left(\frac{4}{8n+1} - \frac{2}{8n+4} - \frac{1}{8n+5} - \frac{1}{8n+6} \right) \left(\frac{1}{16} \right)^n$$

Why is this so special? It allows us to calculate a *specific* digit of π without calculating the preceding digits. The only snag is that these are not decimal digits: they are *hexadecimal* (base 16), from which we can also work out a given digit in base 8 (octal), 4 (quaternary) or 2 (binary). In 1998 Fabrice Ballard used this formula to show that the 100 billionth hexadecimal digit of π

is 9. Within two years, the record had risen to 250 trillion hexadecimal digits (one quadrillion binary digits).

The current record for decimal digits of π is held by Yasumasa Kanada and coworkers, who computed the first 1.2411 trillion digits in 2002.

• •

Infinite Wealth

During the early development of probability theory, a lot of effort was expended – mainly by various members of the Bernoulli family, which had four generations of able mathematicians – on a strange puzzle, the *St Petersburg paradox*.

You play against the bank, tossing a coin until it first lands heads. The longer you keep tossing tails, the more the bank will pay out. In fact, if you toss heads on the first try, the bank pays you £2. If you first toss heads on the second try, the bank pays you £4. If you first toss heads on the third try, the bank pays you £8. In general, if you first toss heads on the nth try, the bank pays you $£2^n$.

The question is: how much should you be willing to pay to take part in the game?

To answer this, you should calculate your 'expected' winnings, in the long run, and the rules of probability tell you how. The probability of a head on the first toss is $\frac{1}{2}$, and you then win £1, so the expected gain on the first toss is $\frac{1}{2} \times 2 = 1$. The probability of a head first arising on the second toss is $\frac{1}{4}$, and you then win £4, so the expected gain on the second toss is $\frac{1}{4} \times 4 = 1$. Continuing in this way, the expected gain on the nth toss is $\frac{1}{2}n \times 2n = 1$. In total, your expected winnings amount to

$$1 + 1 + 1 + 1 + \ldots$$

going on forever, which is infinite. Therefore you should pay the bank an infinite amount to play the game.

What—if anything—is wrong here?

Answer on page 299

• •

Let Fate Decide

Two university mathematics students are trying to decide how to spend their evening.

'We'll toss a coin,' says the first. 'If it's heads, we'll go to the pub for a beer.'

'Great!' says the second. 'If it's tails, we'll go to the movies.'

'Exactly. And if it lands on its edge, we'll study.'

Comment: Twice in my life I have witnessed a coin land on its edge. Once was when I was seventeen, playing a game with some friends, and the coin landed in a groove in the table. The second was in 1997, when I gave the Royal Institution Christmas Lectures on BBC Television. We made a large coin from polystyrene, and a young lady from the audience tossed it in a frying-pan like a pancake. The first time she did so, the coin landed stably on its edge.

Admittedly, it was a rather thick coin.

• •

How Many—

Different sets of bridge hands are there?

$$53,644,737,765,488,792,839,237,440,000$$

if you distinguish hands according to who (N, S, E, W) holds them. If not, divide by 8 (the N–S and E–W pairing has to be maintained) to get

$$6,705,592,220,686,099,104,904,680,000$$

Protons are there in the universe according to Sir Arthur Stanley Eddington?

$$136 \times 2^{256} = 15,747,724,136,275,002,577,605,653,961,181,$$
$$555,468,044,717,914,527,116,709,366,231,425,$$
$$076,185,631,031,296$$

Ways are there to rearrange the first 100 numbers?

> 93,326,215,443,944,152,681,699,238,856,266,700,490,715,
> 968,264,381,621,468,592,963,895,217,599,993,229,915,608,
> 941,463,976,156,518,286,253,697,920,827,223,758,251,185,
> 210,916,864,000,000,000,000,000,000,000,000

unless you argue that 'rearrange' excludes the usual ordering $1, 2, 3, \ldots, 100$. If so, the number is

> 93,326,215,443,944,152,681,699,238,856,266,700,490,715,
> 968,264,381,621,468,592,963,895,217,599,993,229,915,608,
> 941,463,976,156,518,286,253,697,920,827,223,758,251,185,
> 210,916,863,999,999,999,999,999,999,999,999

Zeros are there in a googol?

> 100

Googol is a name invented in 1920 by Milton Sirotta (aged 9), nephew of American mathematician Edward Kasner, who popularised the term in his book *Mathematics and the Imagination*. It is equal to 10^{100}, which is 1 followed by one hundred zeros:

> 10,000,000,000,000,000,000,000,000,000,000,000,000,
> 000,000,000,000,000,000,000,000,000,000,000,000,000,
> 000,000,000,000,000,000

Zeros are there in a googolplex?

> 10^{100}

Googolplex is another invented name, equal to $10^{10^{100}}$, which is 1 followed by 10^{100} zeros. The universe is too small to write it down in full, and the lifetime of the universe is too short anyway. Unless our universe is part of a much larger multiverse, and even then it's hard to see why anyone would bother.

● ●

What Shape is a Rainbow?

Why?

We all remember being told what causes rainbows. Sunlight bounces around inside raindrops, which split the white light into its component colours. Whenever you look directly at a rainbow, the Sun will be behind you, and the rain will be falling in front of you. And to knock it on the head, the teacher showed us how a glass prism splits a ray of white light into all the colours of the rainbow.

A neat piece of misdirection, worthy of a conjurer. That explains the colours. But what about the *shape*?

If it's just a matter of light reflecting back from raindrops, why don't we see the colours wherever the rain is coming down? And if that were happening, wouldn't the colours fuzz out back to white, or maybe a muddy grey? Why is the rainbow a series of coloured arcs? And what shape are the arcs?

Answers on page 300

• •

Alien Abduction

Two aliens from the planet Porqupyne want to abduct two Earthlings, but are blissfully unaware that the objects of their attention are actually pigs. In their formal way, the aliens play a game.

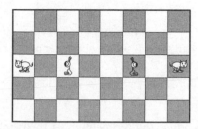

First catch your pig ...

On the first move, each alien moves one square horizontally or vertically (*not* diagonally). Each alien can move in any of the four directions, independently of what the other one does. On the next move, the pigs can do likewise. The aliens get to abduct any pig upon whose square they land. To their surprise, the pigs always seem to get away. *What are the aliens doing wrong?*

Answers on page 302

The Riemann Hypothesis

If there is one single problem that mathematicians would dearly love to solve, it is the Riemann Hypothesis. Entire areas of mathematics would open up if some bright spark could prove this wonderful theorem. And entire areas of mathematics would close down if some bright spark could disprove it. Right now, those areas are in limbo. We can get a glimpse of the Promised Land, but for all we know it might be a mirage.

Oh, there's also a million-dollar prize on offer from the Clay Mathematics Institute.

The story goes back to the time of Gauss, around 1800, and the discovery that although the prime numbers seem rather randomly distributed doing the number line, they have clear *statistical* regularities. Various mathematicians noticed that the number of primes up to some number x, denoted by $\pi(x)$ just to confuse everyone who thought that $\pi = 3.141\,59$, is approximately

$$\pi(x) = \frac{x}{\log x}$$

Gauss found what seemed to be a slightly better approximation, the *logarithmic integral*

$$\mathrm{Li}(x) = \int_2^x \frac{\mathrm{d}x}{\log x}$$

Now, it's one thing to notice this 'prime number theorem', but what counts is proving it, and that turned out to be hard. The most powerful approach is to turn the question into something quite different, in this case complex analysis. The connection between primes and complex functions is not at all obvious, but the key idea was spotted by Euler.

Every positive integer is a product of primes in a unique way. We can formulate this basic property analytically. A first attempt would be to notice that

$$(1 + 2 + 2^2 + 2^3 + \ldots) \times (1 + 3 + 3^2 + 3^3 + \ldots)$$
$$\times (1 + 5 + 5^2 + 5^3 + \ldots) \times \ldots$$

with each bracketed series going on for ever, and taking the product over all primes, is equal to

$$1 + 2 + 3 + 4 + 5 + 6 + 7 + 8 + \ldots$$

summed over all integers. For example, to find out where a number like 360 comes from, we write it as a product of primes

$$360 = 2^3 \times 3^2 \times 5$$

and then pick out from the formula the corresponding terms, here shown in bold:

$$(1 + 2 + 2^2 + \mathbf{2^3} + \ldots) \times (1 + 3 + \mathbf{3^2} + 3^3 + \ldots)$$
$$\times (1 + \mathbf{5} + 5^2 + 5^3 + \ldots) \times \ldots$$

When you 'expand' the brackets, each possible product of prime powers occurs exactly once.

Unfortunately this makes no sense because the series diverge to infinity, and so does the product. However, if we replace each number n by a suitable power n^{-s}, and make s large enough, everything converges. (The minus sign ensures that *large* values of s lead to convergence, which is more convenient.) So we get the formula

$$(1 + 2^{-s} + 2^{-2s} + 2^{-3s} + \ldots) \times (1 + 3^{-s} + 3^{-2s} + 3^{-3s} + \ldots)$$
$$\times (1 + 5^{-s} + 5^{-2s} + 5^{-3s} + \ldots) \times \ldots$$
$$= 1 + 2^{-s} + 3^{-s} + 4^{-s} + 5^{-s} + 6^{-s} + 7^{-s} + 8^{-s} + \ldots$$

(where I've written 1 instead of 1^{-s} because these are equal anyway.) This formula makes perfectly good sense provided s is real and greater than 1. It's true because

$$60^{-s} = 2^{-3s} \times 3^{-2s} \times 5^{-s}$$

and similarly for any positive integer.

In fact, the formula makes perfectly good sense if $s = a+ib$ is complex and its real part a is greater than 1. The final series in the formula is called the *Riemann zeta function* of s, denoted by $\zeta(s)$. Here ζ is the Greek letter zeta.

In 1859 Georg Riemann wrote a brief, astonishingly inventive paper showing that the analytic properties of the zeta function reveal deep statistical features of primes, including Gauss's prime number theorem. In fact, he could do more: he could make the error in the approximation of $\pi(x)$ much smaller, by adding further terms to Gauss's expression. Infinitely many such terms, themselves forming a convergent series, would make the error disappear altogether. Riemann could write down an *exact* expression for $\pi(x)$ as an analytic series.

For the record, here's his formula:

$$\pi(x) + \pi(x^{1/2}) + \pi(x^{1/3}) + \ldots$$
$$= \mathrm{Li}(x) + \int_x^\infty \left[(t^2 - 1)t \log t\right]^{-1} t - \log 2 - \sum_\rho \mathrm{Li}(x^\rho)$$

where ρ runs through the non-trivial zeros of the zeta function. Strictly speaking, this formula is not quite correct when the left-hand side has discontinuities, but that can be fixed up. You can get an even more complicated formula for $\pi(x)$ itself by applying the formula again with x replaced by $x^{1/2}, x^{1/3}$, and so on.

All very pretty, but there was one tiny snag. In order to prove that his series is correct, Riemann needed to establish an apparently straightforward property of the zeta function. Unfortunately, he couldn't find a proof.

All complex analysts learn at their mother's knee (the mother here being Augustin-Louis Cauchy, who along with Gauss first understood the point) that the best way to understand any complex function is to work out where its *zeros* lie. That is: which complex numbers s make $\zeta(s) = 0$? Well, it becomes the best way after some nifty footwork; in the region where the series for $\zeta(s)$ converges, there *aren't* any zeros. However, there is another formula which agrees with the series whenever it converges, but also makes sense when it doesn't. This formula lets us extend the definition of $\zeta(s)$ so that it makes sense for *all* complex numbers s. And this 'analytic continuation' of the zeta function does have zeros. Infinitely many of them.

Some of the zeros are obvious – once you see the formula involved in the continuation process. These 'trivial zeros' are the negative even integers -2, -4, -6, and so on. The other zeros come in pairs $a + ib$ and $a - ib$, and all such zeros that Riemann could find had $a = \frac{1}{2}$. The first three pairs, for instance, are

$$\tfrac{1}{2} \pm 14.13i, \quad \tfrac{1}{2} \pm 21.02i, \quad \tfrac{1}{2} \pm 25.01i$$

Evidence like this led Riemann to conjecture ('hypothesise') that *all* non-trivial zeros of the zeta function must lie on the so-called critical line $\frac{1}{2} + ib$.

If he could prove this statement – the famous *Riemann Hypothesis* – then he could prove that Gauss's approximate formula for $\pi(x)$ is correct. He could improve it to an *exact* – though complicated – formula. Great vistas of number theory would be wide open for development.

But he couldn't, and we still can't.

Eventually, the prime number theorem was proved, independently, by Jacques Hadamard and Charles de la Vallée-Poussin in 1896. They used complex analysis, but managed to find a proof that avoided the Riemann Hypothesis. We now know that the first ten trillion non-trivial zeros of the zeta function lie on the critical line, thanks to Xavier Gourdon and Patrick Demichel in 2004. You might think that ought to settle the matter, but in this area of number theory ten trillion is ridiculously small, and may be misleading.

The Riemann Hypothesis is important for several reasons. If true, it would tell us a lot about the statistical properties of primes. In particular, Helge von Koch proved in 1901 that the Riemann Hypothesis is true if and only if the estimate

$$|\pi(x) - \text{Li}(x)| < C\sqrt{x}\log x$$

for the error in Gauss's formula holds for some constant C. Later, Lowell Schoenfeld proved that we can take $C = 1/8\pi$ for all $x \geqslant 2,657$. (Sorry, this area of mathematics does that kind of thing.) The point here is that the error is small compared with x, and it tells us how much the primes fluctuate away from their more typical behaviour.

Riemann's exact formula, of course, would also follow from the Riemann Hypothesis. So would a huge list of other mathematical results – you can find some of them at en.wikipedia.org/wiki/Riemann_hypothesis

However, the main reason why the Riemann Hypothesis is important – apart from 'because it's there' – is that it has a lot of far-reaching analogues and generalisations in algebraic number theory. A few of the analogues have even been proved. There is a feeling that if the Riemann Hypothesis can be proved in its original form, then so can the generalisations. These ideas are too technical to describe, but see mathworld.wolfram.com/RiemannHypothesis.html

I will tell you one deceptively simple statement that is equivalent to the Riemann Hypothesis. Of itself, it looks harmless and unimportant. Not so! Here's how it goes. If n is a whole number, then the sum of its divisors, including n itself, is written as $\sigma(n)$. (Here σ is the lower-case Greek letter 'sigma'.) So

$$\sigma(24) = 1 + 2 + 3 + 4 + 6 + 8 + 12 + 24 = 60$$
$$\sigma(12) = 1 + 2 + 3 + 4 + 6 + 12 = 28$$

and so on. In 2002 Jeffrey Lagarias proved that the Riemann Hypothesis is equivalent to the inequality

$$\sigma(n) \leqslant e^{H_n} \log H_n$$

for every n. Here H_n is the nth harmonic number, equal to

$$1 + \frac{1}{2} + \frac{1}{3} + \frac{1}{4} + \dots \frac{1}{n}$$

• •

Anti-Atheism

Godfrey Harold Hardy, a Cambridge mathematician who worked mainly in analysis, claimed to believe in God – but unlike most believers, he considered the Deity to be his personal enemy. Hardy had it in for God, and he was convinced that God had it in for Hardy, which was only fair. Hardy was especially worried whenever he travelled by sea, in case God sank the boat. So before travelling, he would send his colleagues a telegram: 'HAVE PROVED RIEMANN HYPOTHESIS. HARDY.' He would then retract this claim on arrival.

As just discussed, the Riemann Hypothesis is the most famous unsolved problem in mathematics, and one of the most important. And so it was in Hardy's day, too. When his colleagues asked him why he sent such telegrams, he explained that God would never let him die if that would give him credit – however controversial – for proving the Riemann Hypothesis.

• •

Disproof of the Riemann Hypothesis

Consider the following logical argument:

- Elephants never forget.
- No creature that has ever won *Mastermind* has possessed a trunk.
- A creature that never forgets will always win *Mastermind* provided it takes part in the competition.
- A creature lacking a trunk is not an elephant.
- In 2001 an elephant took part in *Mastermind*.

Therefore:

- The Riemann Hypothesis is false.

Is this a correct deduction?

  Answer on page 302

• •

Murder in the Park

This puzzle – like several in this book – goes back to the great English puzzlist Henry Ernest Dudeney. He called it 'Ravensdene Park'. I've made a few trivial changes.

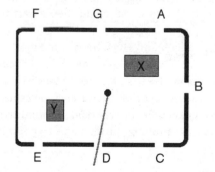

Ravensdene Park.

  Soon after a heavy fall of snow, Cyril Hastings entered Ravensdene Park at gate D, walked straight to the position

marked with a black dot, and was stabbed through the heart. His body was found the next morning, along with several tracks in the snow. The police immediately closed the park.

Their subsequent investigations revealed that each track had been made by a different, very distinctive shoe. Witnesses placed four individuals, other than Hastings, in the park during the period concerned. So the murderer had to be one of them. Examining their shoes, the police deduced that:

- The butler – who could prove he had been in the house X at the time of the murder – had entered at gate E and gone to X.
- The gamekeeper – who had no such alibi – had entered at gate A and gone to his lodge at Y.
- A local youth had entered at gate G and left by gate B.
- The grocer's wife had entered at gate C and left by gate F.

None of these individuals entered or left the park more than once.

It had been foggy as well as snowy, so the routes these people took were often rather indirect. The police did notice that no two paths crossed. But they failed to make a sketch of the routes before the snow melted and they disappeared.

So who was the murderer?

Answer on page 303

• •

The Cube of Cheese

An oldie, but none the worse for that. Marigold Mouse has a cube of cheese and a carving-knife. She wishes to slice the cheese along a flat plane, to obtain a cross-section that is a regular hexagon. Can she do this, and if so, how?

Answer on page 304

• •

The Game of Life

In the 1970s John Conway invented the Game of Life. Strange black creatures scuttle across a grid of white tiles, changing shape, growing, collapsing, freezing and dying. The best way to play Life (as it's commonly known) is to download suitable software. There are several excellent free programs for Life on the web, easily located by searching. A Java version, which is easy to use and will give hours of pleasure, can be found at: www.bitstorm.org/gameoflife/

Life is played with black counters on a potentially infinite grid of square cells. Each cell holds either one counter or none. At each stage, or *generation*, the set of counters defines a *configuration*. The initial configuration at generation 0 evolves at successive stages according to a short list of rules. The rules are illustrated in the diagrams below. The *neighbours* of a given cell are the eight cells immediately adjacent to it, horizontally, vertically or diagonally. All births and deaths occur simultaneously: what happens to each counter or empty cell in generation $n+1$ depends only on its neighbours in generation n.

An occupied cell and its eight neighbours.

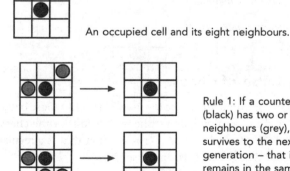

current generation   next generation

Rule 1: If a counter (black) has two or three neighbours (grey), it survives to the next generation – that is, it remains in the same cell.

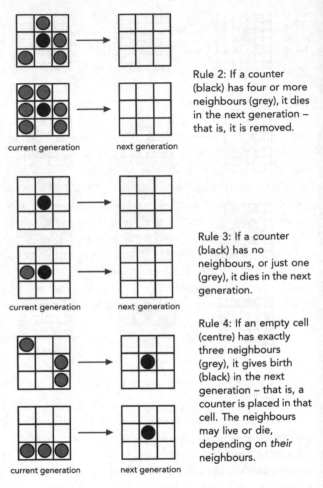

current generation   next generation

Rule 2: If a counter (black) has four or more neighbours (grey), it dies in the next generation – that is, it is removed.

Rule 3: If a counter (black) has no neighbours, or just one (grey), it dies in the next generation.

Rule 4: If an empty cell (centre) has exactly three neighbours (grey), it gives birth (black) in the next generation – that is, a counter is placed in that cell. The neighbours may live or die, depending on *their* neighbours.

Starting from any given initial configuration, the rules are applied repeatedly to produce the life history of its succeeding generations. For example, here is the life history of a small triangle built from four counters:

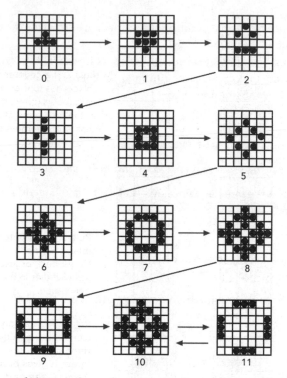

Life history of a configuration. Configurations 8 and 9 alternate periodically.

Even this simple examples shows that the rules for Life can generate complex structures from simpler ones. Here the sequence of generations becomes *periodic*: at generation 10 the configuration is the same as at generation 8, and thereafter configurations 8 and 9 alternate, a sequence known as *traffic lights*.

One of the fascinations of Life is the astonishing variety of life histories, and the absence of any *obvious* relation between the initial configuration and what it turns into. The system of rules is entirely deterministic – the entire infinite future of the system is implicit in its initial state. But Life dramatically demonstrates the difference between determinism and predictability. This is where the name 'Life' comes from.

From a mathematical point of view, it is natural to classify Life configurations according to their long-term behaviour. For example, configurations may:

(1) disappear completely (die)

(2) attain a steady state (stasis)

(3) repeat the same sequence over and over again (periodicity)

(4) repeat the same sequence over and over again but end up in a new location

(5) behave chaotically

(6) exhibit computational behaviour (universal Turing machine).

Among the common periodic configurations are the *blinker* and traffic lights, with periods 3 and 8, respectively:

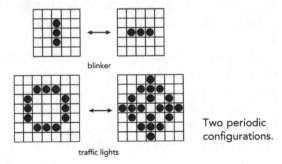

blinker

traffic lights

Two periodic configurations.

The outcome of a game of Life is extraordinarily sensitive to the precise choice of the initial state. A difference of one cell can totally change the state's future. Moreover, simple initial configurations can sometimes develop into very complicated ones. This behaviour – to some extent 'designed in' by the choice of update rules – motivates the game's name.

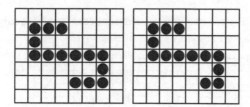

The left-hand S-shaped configuration eventually settles down after 1,405 generations, by which time it has given birth to 2 gliders, 24 blocks, 6 ponds, 4 loaves, 18 beehives and 8 blinkers. If you delete just one cell to give the right-hand configuration, everything dies completely after 61 generations.

The prototype mobile state is the *glider*, which moves one cell diagonally every four moves:

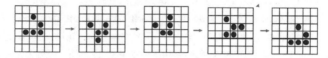

Motion of a glider.

Three *spaceships* (lightweight, middleweight, heavyweight) repeat cycles that cause them to move horizontally, throwing off sparks that vanish immediately. Longer spaceships do not work on their own – they break apart in complicated ways – but they can be supported by flotillas of accompanying smaller spaceships.

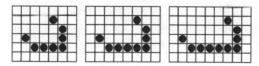

Spaceships.

One of the earliest mathematical questions about Life was whether there can exist a finite initial configuration whose future configurations are *unbounded* – that is, it will become as large as we wish if we allow enough time to pass. This question was answered in the affirmative by Bill Gosper's invention of a *glider gun*. The configuration shown below in black oscillates with period 30, and repeatedly fires gliders (the first two are shown in grey). The stream of gliders grows unboundedly.

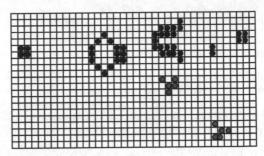

The glider gun, with two gliders it has spat out.

It turns out that the Game of Life has configurations which act like computers, able in principle to calculate anything that a computer program can specify. For example, such a configuration can compute π to as many decimal places as we want. In practice, such computations run incredibly slowly, so don't throw away your PC just yet.

In fact, even simpler 'games' of the same kind – known as cellular automata (page 239) – living on a line of squares instead of a two-dimensional array, can behave as universal computers. This automaton, known as 'rule 110', was suggested by Stephen Wolfram in the 1980s, and Matthew Cook proved its universality in the 1990s. It illustrates, in a very dramatic way, how astonishingly complex behaviour can be generated by very simple rules. See mathworld.wolfram.com/Rule110.html

. . . .

Two-Horse Race

Every whole number can be obtained by multiplying suitable primes together. If this requires an even number of primes, we say that the number is of *even type*. If it requires an odd number of primes, we say that the number is of *odd type*. For instance,

$$96 = 2 \times 2 \times 2 \times 2 \times 2 \times 3$$

uses six primes, so is of even type. On the other hand,

$$105 = 3 \times 5 \times 7$$

uses three primes, so is of odd type. By convention, 1 is of even type.

For the first ten whole numbers, 1–10, the types are:

Odd		2	3		5		7	8		
Even	1			4		6			9	10

A striking fact emerges: in general, odd types occur at least as frequently as even types. Imagine two horses, Odd and Even, racing. Start them level with each other, and read along the sequence of numbers: $1, 2, 3, \ldots$ At each stage, move Odd forward one step if the next number has odd type; move Even forward one step if the next number has even type. So:

> After 1 step, Even is ahead.
> After 2 steps, Odd and Even are level.
> After 3 steps, Odd is ahead.
> After 4 steps, Odd and Even are level.
> After 5 steps, Odd is ahead.
> After 6 steps, Odd and Even are level.
> After 7 steps, Odd is ahead.
> After 8 steps, Odd is ahead.
> After 9 steps, Odd is ahead.
> After 10 steps, Odd and Even are level.

Odd always seems to be level, or ahead. In 1919 George Pólya conjectured that Odd *never* falls behind Even, except right at the start, step 1. Calculations showed that this is true for the first million steps. Given this weight of favourable evidence, surely it has to be true for any number of steps?

Without a computer you can waste a lot of time on this question, so I'll tell you the answer. Pólya was *wrong*! In 1958, Brian Haselgrove proved that at some (unknown) stage Odd falls behind Even. Once reasonably fast computers were available, it was easy to test ever larger numbers. In 1960 Robert Lehman discovered that Even is in the lead at step 906,180,359. In 1980

Minoru Tanaka proved that Even *first* takes the lead at step 906,150,257.

This kind of thing is what makes mathematicians insist upon proofs. And it shows that even a number like 906,150,257 can be interesting and unusual.

• •

Drawing an Ellipse – and More?

It is well known that an easy way to draw an ellipse is to fix two pins through the paper, tie a loop of string round them, and place your pencil so that the string stays taut. Gardeners sometimes use this method to map out elliptical flowerbeds. The two pins are the *foci* (plural of *focus*, and pronounced 'foe-sigh') of the ellipse.

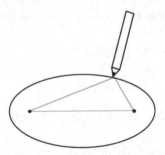

How to draw an ellipse.

Suppose that you use three pegs, in a triangle. It need not be an equilateral or isosceles triangle.

Why isn't this interesting too?

That ought to give some interesting new kinds of curves. So why don't the mathematics books mention them?

Answer on page 304

. . . .

Mathematical Jokes 3

Two mathematicians in a cocktail bar are arguing about how much maths the ordinary person knows. One thinks they're hopelessly ignorant; the other says that quite a few people actually know a lot about the subject.

'Bet you twenty pounds I'm right,' says the first, as he heads for the gents. While he is gone, his colleague calls the waitress over.

'Listen, there's ten pounds in it for you if you come over when my friend gets back and answer a question. The answer is "one-third x cubed." Got that?'

'Ten pounds for saying "One thirdex cue?"'

'No, one-third x cubed.'

'One thir dex cubed?'

'Yeah, that'll do.'

The other mathematician comes back, and the waitress comes over.

'Hey – what's the integral of x squared?'

'One third x cubed,' says the waitress. As she walks away, she adds, over her shoulder, 'Plus a constant.'

. . . .

The Kepler Problem

Mathematicians have learned that apparently simple questions are often hard to answer, and apparently obvious facts may be false, or may be true but extremely hard to prove. The Kepler problem is a case in point: it took nearly three hundred years to solve it, even though everyone knew the correct answer from the start.

It all began in 1611 when Johannes Kepler, a mathematician and astrologer (yes, he cast horoscopes; lots of mathematicians did at that time – it was a quick way to make money) wanted to give his sponsor a New Year's gift. The sponsor rejoiced in the name Johannes Mathäus Wacker of Wackenfels, and Kepler wanted to say 'thanks for all the cash' without actually spending any of it. So he wrote a book, and presented it to his sponsor. Its title (in Latin) was *The Six-Cornered Snowflake*. Kepler started with the curious shapes of snowflakes, which often form beautiful sixfold symmetric crystals, and asked why this happened.

A typical 'dendritic' snowflake.

It is often said that 'no two snowflakes are alike'. The logician in me objects 'How can you tell?' but a back-of-the-envelope calculation suggests that there are so many features in a 'dendritic' snowflake, of the kind illustrated, that the chance of two being identical is pretty much zero.

No matter. What matters here is that Kepler's analysis of the snowflake led him to the idea that its sixfold symmetry arises because that's the most efficient way to pack circles in a plane.

Take a lot of coins, of the same denomination – pennies, say. If you lay them on a table and push them together tightly, you quickly discover that they fit perfectly into a honeycomb pattern, or 'hexagonal lattice':

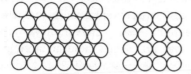

(Left) The closest way to pack circles, and (right) a less efficient lattice packing.

And this is the *closest* packing – the one that fills space most efficiently, in the ideal case of infinitely many circles arranged on a plane. Alternatives, such as the square lattice on the right, are less efficient.

Mind you, this innocent assertion wasn't proved until 1940, when László Fejes Tóth managed it. (Axel Thue sketched out a proof in 1892, and gave more details in 1910, but he left some gaps.) Tóth's proof was quite hard. Why the difficulty? We don't know, to begin with, that the most efficient packing forms a regular lattice. Maybe something more random could work better. (For *finite* packings, say inside a square, this can actually happen – see the next puzzle, about a milk crate.)

Along the way, Kepler came very close indeed to the idea that all matter is made from tiny indivisible components, which we now call 'atoms'. This is impressive, given that he did no experiments in the course of writing his book. Atomic theory, introduced by the Greek Democritus, was not established experimentally until about 1900.

Kepler had his eye on something a bit more complicated, though: the closest way to pack identical spheres in space. He was aware of three regular 'lattice' packings, which we now call the *hexagonal*, *cubic* and *face-centred cubic* lattices. The first of these is formed by stacking lots of honeycomb layers of spheres on top of one another, with the centres of corresponding spheres forming a vertical line. The second is made from square-lattice layers, also stacked vertically. For the third, we stack hexagonal layers, but fit the spheres in any given layer into the hollows in the one below.

You can get the same result, though tilted, by similarly stacking square-lattice layers so that the spheres in any given layer fit into the hollows in the one below – this isn't entirely obvious, and – like the milk crate puzzle – it shows that intuition may not be a good guide in this area. The picture shows how this happens: the horizontal layers are square, but the slanting layers are hexagonal.

Part of a face-centred cubic lattice.

Now, every greengrocer knows that the way to stack oranges is to use the face-centred cubic lattice.* By thinking about pomegranate seeds, Kepler was led to the casual remark that with the face-centred cubic lattice, 'the packing will be the tightest possible'.

That was in 1611. The proof that Kepler was right had to wait until 1998, when Thomas Hales announced that he had achieved this with massive computer assistance. Basically, Hales considered all possible ways to surround a sphere with other spheres, and showed that if the arrangement wasn't the one found in the face-centred cubic lattice, then the spheres could be shoved closer together. Tóth's proof in the plane used the same ideas, but he only had to check about forty cases.

Hales had to check thousands, so he rephrased the problem in terms that could be verified by a computer. This led to a huge computation – but each step in it is essentially trivial. Almost all of the proof has been checked independently, but a very tiny level of doubt still remains. So Hales has started a new computer-

---

* They don't say it that way, but they stack it that way.

based project to devise a proof that can be verified by standard proof-checking software. Even then, a computer will be involved in the verifications, but the software concerned does such simple things – in principle – that a human can check that the software does what it is supposed to. The project will probably take 20 years. You can still object on philosophical grounds if you wish, but you'll be splitting logical hairs very finely indeed.

What makes the problem so hard? Greengrocers usually start with a square box that has a flat base, so they naturally pack their oranges in layers, making each layer a square lattice. It is then natural to make the second layer fill the gaps in the bottom one, and so on. If by chance they start with a hexagonal layer instead, they get the same packing anyway, except for a tilt. Gauss proved in 1831 that Kepler's packing is the tightest *lattice* packing. But the mathematical problem here is to prove this, without assuming at the start that the packing forms flat layers. The mathematician's spheres can hover unsupported in space. So the greengrocer's 'solution' involves a whole pile of assumptions – well, actually, oranges. Since experiments aren't proofs, and here even the experiment is dodgy, you can see that the problem could be harder than it seems

. . . .

The Milk Crate Problem

Here's a simpler question of the same kind. A milkman wishes to pack identical bottles, with circular cross-section, into a square crate. To him, it's obvious that for any given *square* number of bottles – 1, 4, 9, 16, and so on – the crate can be made as small as possible by packing the bottles in a regular square array. (He can see that with a non-square number of bottles, there are gaps and maybe the bottles can be jiggled around to shrink the crate.)

Is he right?

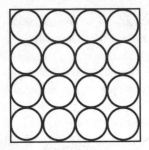

How the milkman fits
16 bottles into the
smallest possible
square crate.

Answer on page 305

• •

Equal Rights

One of the leading female mathematicians of the early twentieth
century was Emmy Noether, who studied at the University of
Göttingen. But after she completed her doctorate the authorities
refused to allow her to proceed to the status of *Privatdozent*,
which would allow her to charge students fees for tuition. Their
stated reason was that women were not permitted to attend
faculty meetings at the university senate. The head of the
mathematics department, the great David Hilbert, is said to have
remarked: 'Gentlemen! There is nothing wrong with having a
woman in the senate. Senate is not a public bath.'

• •

Road Network

Four towns – Aylesbury, Beelsbury, Ceilsbury and Dealsbury – lie
at the corners of a square, of side 100 km. The highways
department wishes to connect them all together using the
shortest possible network of roads.

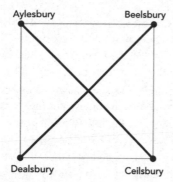

Not like this.

'We can run roads straight from Aylesbury to Beelsbury to Ceilsbury to Dealsbury,' said the assistant town planner. 'That's 300 km of roads.'

'No, we can do better than that!' his boss replied. 'Two diagonals, which, if you recall your Pythagoras, amount to 200 √2 km – about 282 km.'

What is the shortest network? Using the diagonals of the square is *not* the answer.

Answer on page 305

Answer on page 305

• •

Tautoverbs

In Terry Pratchett's *Discworld* series of fantasy novels, the members of the Order of Wen the Eternally Surprised, better known as the History Monks, are greatly impressed by the homespun wisdom of Mrs Marietta Cosmopilite. They have never before heard her homely homilies (such as 'I haven't got all day, you know'), so to those monks who follow the Way of Mrs Cosmopilite, her simple assertions are marvellous new philosophical insights.

Mathematicians take a more jaundiced view of folk wisdom, and habitually revise proverbs to make them more logical. Indeed tautological – trivially true. Thus the proverb 'Penny wise, pound foolish' becomes the *tautoverb* 'Penny wise, wise about pennies,' which makes more sense and is difficult to dispute. And 'Look after the pennies and the pounds will look after themselves' is more convincing in the form 'Look after the pennies and the pennies will be looked after.'

As a kid I was always vaguely bothered that in their original form these two proverbs contradict each other, though I now see this as the default mechanism whereby folk wisdom ensures that it is perceived to be wise. The revised versions do *not* conflict – clear evidence of their superiority. I'll give you a couple of other examples to get you started, and then turn you lose on the opening words of several proverbs. Your job is to complete them to make tautoverbs. The first example is simple and direct, the other more baroque. Both forms are permissible. So is helpful commentary, preferably blindingly obvious. Logical quibbles are actively encouraged – the more pedantic, the better.

- He who fights and runs away will live to run away again.
- A bird in the bush is worth two in the hand, because free-range produce is always expensive.

OK, now it's your turn. In the same spirit, complete the following tautoverbs:

- No news is—
- The bigger they are—
- Nothing ventured—
- Too many cooks—
- You cannot have your cake—
- A watched pot—
- If pigs had wings—

If you have enjoyed this game, psychiatric help is recommended, but until it arrives you can find a lot more proverbs to work on at www.manythings.org/proverbs/index.html

Answers on page 306

. . . .

Complexity Science

Complexity science, or the theory of complex systems, came to prominence with the founding of the Santa Fe Institute (SFI) in 1984, by George Cowan and Murray Gell-Mann. This was, and still is, a private research centre for interdisciplinary science, with emphasis on 'the sciences of complexity'. You might think that 'complexity' refers to anything complicated, but the SFI's main objective has been to develop and disseminate new mathematical techniques that could shed light on systems in which very large numbers of agents or entities interact with one another according to relatively simple rules. A key phenomenon is what is called *emergence*, in which the system as a whole behaves in ways that are not available to the individual entities.

An example of a real-world complex system is the human brain. Here the entities are nerve cells – neurons – and the emergent features include intelligence and consciousness. Neurons are neither intelligent nor conscious, but when enough of them are hooked together, these abilities emerge. Another example is the world's financial system. Now the entities are bankers and traders, and emergent features include stock-market booms and crashes. Other examples are ants' nests, ecosystems and evolution. You can probably work out what the entities are for each of these, and think of some emergent features. Anyone can play this game.

What's harder, and what SFI was, and still is, all about, is to model such systems mathematically in a way that reflects their underlying structure as an interacting system of simple components. One such modelling technique is to employ a cellular automaton – a more general version of John Conway's Game of Life. This is like a computer game played on a square grid. At any given instant, each square exists in some state, usually represented by what colour it is. As time ticks to the next instant, each square changes colour according to some list of rules. The rules involve the colours of neighbouring squares, and might be something like this: 'a red square changes to green if it has between two and six blue neighbours'. Or whatever.

Three types of pattern formed by a simple cellular automaton: static (blocks of the same colour), structured (the spirals), and chaotic (for example the irregular patch at bottom right).

It might seem unlikely that such a rudimentary gadget can achieve anything interesting, let alone solve deep problems of complexity science, but it turns out that cellular automata can behave in rich and unexpected ways. In fact their earliest use, by John von Neumann in the 1940s, was to prove the existence of an abstract mathematical system that could self-replicate – make copies of itself.* This suggested that the ability of living creatures to reproduce is a logical consequence of their physical structure, rather than some miraculous or supernatural process.

Evolution, in Darwin's sense, offers a typical example of the complexity-theory approach. The traditional mathematical model of evolution is known as population genetics, which goes

---

* There is now a lot of interest in doing the same with real machines, using nanotechnology. There are many science fiction stories about 'Von Neumann machines', often employed by aliens or machine cultures to invade planets, including our own. The techniques used to pack millions of electronic components on to a tiny silicon chip are now being used to build extremely tiny machines, 'nanobots', and a true replicating machine may not be so far away. Alien invasions are not a current cause for concern, but the possibility of a mutant Von Neumann machine turning the Earth into 'grey goo' has raised issues about the safety, and control, of nanotechnology. See en.wikipedia.org/wiki/Grey_goo

back to the British statistician Sir Ronald Fisher, around 1930.
This approach views an ecosystem – a rainforest full of different
plants and insects, or a coral reef—as a vast pool of genes. As the
organisms reproduce, their genes are mixed together in new
combinations.

For example, a hypothetical population of slugs might have
genes for green or red skins, and other genes for a tendency to
live in bushes or on bright red flowers. Typical gene combinations are green–bush, green–flower, red–bush, and red–flower.
Some combinations have greater survival value than others. For
example red–bush slugs would easily be seen by birds against the
green bushes they live in, whereas red–flower slugs would be less
visible.

As natural selection weeds out unfit combinations, the
combinations that allow organisms to survive better tend to
proliferate. Random genetic mutations keep the gene pool
simmering. The mathematics centres on the *proportions* of
particular genes in the population, and works out how those
proportions change in response to selection.

A complexity model of slug evolution would be very
different. For instance, we could set up a cellular automaton,
assigning various environmental characteristics to each cell. For
example, a cell might correspond to a piece of bush, or a flower,
or whatever. Then we choose a random selection of cells and
populate them with 'virtual slugs', assigning a combination of
slug genes to each such cell.

Other cells could be 'virtual predators'. Then we specify rules
for how the virtual organisms move about the grid and interact
with one another. For example, at each time-step a slug must
either stay put or move to a random neighbouring cell. On the
other hand, a predator might 'see' the nearest slug and move five
cells towards it, 'eating' it if it reaches the slug's own cell – so that
particular virtual slug is removed from the computer's memory.

We would set up the rules so that green slugs are less likely to
be 'seen' if they are on bushes rather than flowers. Then this
mathematical computer game would be allowed to run for a few

million time-steps, and we would read off the proportions of various surviving slug gene combinations.

Complexity theorists have invented innumerable models in the same spirit: building in simple rules for interactions between many individuals, and then simulating them on a computer to see what happens. The term 'artificial life' has been coined to describe such activities. A celebrated example is Tierra, invented by Tom Ray around 1990. Here, short segments of computer code compete with one another inside the computer's memory, reproducing and mutating (see www.nis.atr.jp/~ray/tierra/). His simulations show spontaneous increases in complexity, rudimentary forms of symbiosis and parasitism, lengthy periods of stasis punctuated by rapid changes – even a kind of sexual reproduction. So the message from the simulations is that all these puzzling phenomena are entirely natural, provided they are seen as emergent properties of simple mathematical rules.

The same difference in working philosophy can be seen in economics. Conventional mathematical economics is based on a model in which every player has complete and instant information. As the Stanford economist Brian Arthur puts it, the assumption is that 'If two businessmen sit down to negotiate a deal, in theory each can instantly foresee all contingencies, work out all possible ramifications, and effortlessly choose the best strategy.' The goal is to demonstrate mathematically that any economic system will rapidly home in on an equilibrium state, and remain there. In equilibrium, every player is assured of the best possible financial return for themselves, subject to the overall constraints of the system. The theory puts formal flesh on the verbal bones of Adam Smith's 'invisible hand of the market'.

Complexity theory challenges this cosy capitalist utopia in a number of ways. One central tenet of classical economic theory is the 'law of diminishing returns', which originated with the English economist David Ricardo around 1820. This law asserts that any economic activity that undergoes growth must eventually be limited by constraints. For example, the plastics industry depends upon a supply of oil as raw material. When oil is cheap, many companies can move over from, say, metal components to plastic ones. But this creates increasing demand for oil, so the price goes up. At some level, everything balances out.

Modern hi-tech industries, however, do not follow this pattern at all. It costs perhaps a billion dollars to set up a factory to make the latest generation of computer memory chips, and until the factory begins production, the returns are zero. But once the factory is in operation, the cost of producing chips is tiny. The longer the production run, the cheaper chips are to make. So here we see a law of *increasing* returns: the more goods you make, the less it costs you to do so.

From the point of view of complex systems, the market is not a simple mathematical equilibrium-seeker, but a 'complex adaptive system', where interacting agents modify the rules that govern their own behaviour. Complex adaptive systems often settle into interesting patterns, strangely reminiscent of the complexities of the real world. For example, Brian Arthur and his colleagues have set up computer models of the stock market in which the agents search for patterns – genuine or illusory – in the market's behaviour, and adapt their buying and selling rules according to what they perceive. This model shares many features of real stock markets. For example, if many agents 'believe' that the price of a stock will rise, they buy it, and the belief becomes self-fulfilling.

According to conventional economic theory, none of these phenomena should occur. So why do they happen in complexity models? The answer is that the classical models have inbuilt mathematical limitations, which preclude most kinds of 'interesting' dynamics. The greatest strength of complexity theory is that it resembles the untidy creativity of the real world. Paradoxically, it makes a virtue of simplicity, and draws far-ranging conclusions from models with simple – but carefully chosen – ingredients.

● ●

Scrabble Oddity

The letter scores at Scrabble are:

Score 1	A, E, I, L, N, O, R, S, T, U
Score 2	D, G
Score 3	B, C, M, P
Score 4	F, H, V, W, Y
Score 5	K
Score 8	J, X
Score 10	Q, Z

Which positive integer is equal to its own Scrabble score when spelt out in full?

Answer on page 306

• •

Dragon Curve

The picture shows a sequence of curves, called dragon curves (look at the last one). The sequence can be continued indefinitely, getting ever more complicated curves.

What is the rule for making them? Ignore the 'rounding' of the corners by the short lines, which is done so that later curves in the sequence remain intelligible.

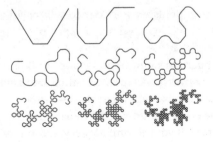

The first
nine
dragon curves.

Answer on page 307

• •

Counterflip

Make some circular counters out of card, each black on one side and white on the other (the precise number doesn't matter, but 10 or 12 is about right). Arrange them in a row, with a random choice of colours facing upwards.

Your task is now to remove all the counters, by making a series of moves. Each move involves choosing a black counter, removing it, and flipping any neighbouring counters over to change their colours. Counters are 'neighbours' if they are next to each other in the original row of counters; removing any counter creates a gap. As the game progresses, a counter may have two, one or no neighbours.

Here is a sample game in which the player succeeds in removing all the counters:

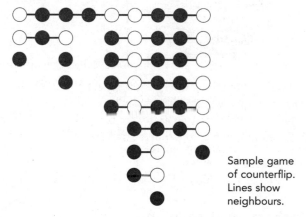

Sample game of counterflip. Lines show neighbours.

The key to this puzzle is simple, but far from obvious: with correct play, you can always succeed if the initial number of black counters is odd. If it is even, there is no solution.

You can play the game for fun, without analysing its mathematical structure. If you feel ambitious, you can look for a winning strategy – and explain why there is no way to win when the initial number of black counters is even.

Answer on page 307

Answer on page 307

Spherical Sliced Bread

Araminta Ponsonby took her two sets of quins to the Archimedes bakery, which makes spherical loaves. She likes to go there because each loaf is cut into ten slices, of equal thickness, so each child can have one slice of bread. They have different appetites – which is fortunate because some slices have smaller volume than others. But, being extremely well-behaved, all ten children love the crust, and want as much as they can get.

Which slice has the most crust?

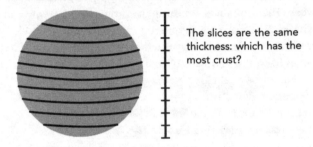

The slices are the same thickness: which has the most crust?

Assume that the loaf is a perfect sphere, the slices are formed by parallel equally spaced planes, and the crust is infinitely thin – so the amount of crust on each slice is equal to the *area* of the corresponding part of the sphere's surface.

Answer on page 309

• •

Mathematical Theology

It is said that during Leonhard Euler's second stint at the Court of Catherine the Great, the French philosopher Denis Diderot was trying to convert the Court to atheism. Since royalty generally claims to have been appointed by God, this didn't go down terribly well. At any rate, Catherine asked Euler to put a spoke in Diderot's wheel. So Euler told the Court that he knew an algebraic proof of the existence of God. Facing Diderot, he declaimed: 'Sir,

$a + b^n/n = x$, hence God exists – reply!' Diderot had no answer, and left the Court to widespread laughter, humiliated.

Yes, well ... There are some little problems with this anecdote, which seems to have originated with the English mathematician Augustus De Morgan in his *Budget of Paradoxes*. As the historian Dirk Struik pointed out in 1967, Diderot was an accomplished mathematician who had published work on geometry and probability, and would have been able to recognise nonsense when he heard it. Euler, an even better mathematician, would not have expected something that simple-minded to work. The formula is a meaningless equation unless we know what a, b, n and x are supposed to be. As Struik remarks, 'No reason exists to think that the thoughtful Euler would have behaved in the asinine way indicated.'

Euler was a religious man, who apparently considered the Bible to be literal truth, but he also believed that knowledge stems, in part, from rational laws. In the eighteenth century there was occasional talk about the possibility of an algebraic proof of the Deity's existence, and Voltaire mentions one by Maupertuis in his *Diatribe*.

A much better attempt was found among Kurt Gödel's unpublished papers. Naturally, it is formulated in terms of mathematical logic, and for the record here it is in its entirety:

$Ax.1$ $\Box \forall x[\phi(x) \to \psi(x)] \wedge P(\phi) \to P(\psi)$
$Ax.2$ $P(\neg\phi) \sqrt{} \neg P(\phi)$
$Th.1$ $P(\phi) \to \Diamond \exists x[\phi(x)]$
$Df.1$ $G(x) \Leftrightarrow \forall\phi[P(\phi) \to \phi(x)]$
$Ax.3$ $P(G)$
$Th.2$ $\Diamond \exists x G(x)$
$Df.2$ $\phi \,\mathrm{ess}\, x \Leftrightarrow \phi(x) \wedge \forall\Psi\Psi(x) \to \Box\forall x[\phi(x) \to \Psi(x)]$
$Ax.4$ $P(\phi) \to \Box P(\phi)$
$Th.3$ $G(x) \to G \,\mathrm{ess}\, x$
$Df.3$ $E(x) \Leftrightarrow \forall\phi[\phi \,\mathrm{ess}\, x \to \Box \exists x \phi(x)$
$Ax.5$ $P(E)$
$Th.4$ $\Box \exists x G(x)$

The symbolism belongs to a branch of mathematical logic called *modal logic*. Roughly speaking, the proof works with 'positive properties', denoted by P. The expression $P(\phi)$ means that ϕ is a positive property. The property 'being God' is defined (Df.1) by requiring God to have *all* positive properties. Here $G(x)$ means 'x has the property of being God', which is a fancy way of saying 'x is God'. The symbols $\square$ and $\Diamond$ denote 'necessary truth,' and 'contingent truth,' respectively. The arrow $\rightarrow$ means 'implies', $\forall$ is 'for all' and $\exists$ is 'there exists'. The symbol $\neg$ means 'not', $\wedge$ is 'and', and $\leftrightarrow$ and $\Leftrightarrow$ are subtly different versions of 'if and only if'. The symbol 'ess' is defined in Df.2. The axioms are Ax.1–5. The theorems (Th.1–4) culminate in the statement 'there exists x such that x has the property of being God' – that is, God exists.

The distinction between necessary and contingent truth is a key novelty of modal logic. It distinguishes statements that *must* be true (such as '$2 + 2 = 4$' in a suitable axiomatic treatment of mathematics) from those that conceivably might be false (such as 'it is raining today'). In conventional mathematical logic, the statement 'If A then B' is always considered to be true when A is false. For instance '$2 + 2 = 5$ implies $1 = 1$' is true, and so is '$2 + 2 = 5$ implies $1 = 42$'. This may seem strange, but it is possible to prove that $1 = 1$ starting from $2 + 2 = 5$, and it is also possible to prove that $1 = 42$ starting from $2 + 2 = 5$. So the convention makes good sense. *Can you find any such proofs?*

If we extend this convention to human activities, then the statement 'If Hitler had won World War II then Europe would now be a single nation' is trivially true, because Hitler did *not* win World War II. But 'If Hitler had won World War II then pigs would now have wings' is *also* trivially true, for the same reason. In modal logic, however, it would be sensible to debate the truth or falsity of the first of these statements, depending on how history might have changed if the Nazis had won the war. The second would be false, because pigs don't have wings.

Gödel's sequence of statements turns out to be a formal version of the ontological argument put forward by St Anselm of Canterbury in his *Proslogion* of 1077–78. Defining 'God' as 'the greatest conceivable entity', Anselm argued that God is conceivable. But if he is not real, we could conceive of Him being greater by existing in reality. Therefore, God must be real.

Aside from deep issues of what we mean by 'greatest' and so forth, there is a basic logical flaw here, one that every mathematician learns at his mother's knee. Before we can deduce any property of some entity or concept from its definition, we must first prove that something satisfying the definition exists. Otherwise the definition might be self-contradictory. For instance, suppose we define n to be 'the largest whole number'. Then we can easily prove that $n = 1$. For if not, $n^2 > n$, contradicting the definition of n. Therefore 1 is the largest whole number. The flaw is that we cannot use any properties of n until we know that n exists. As it happens, it doesn't – but even if it did, we would have to *prove* that it did before proceeding with the deduction.

In short: in order to prove that God exists by Anselm's line of thinking, we must first establish that God exists (by some other line of reasoning, or else the logic is circular). Of course I've simplified things here, and later philosophers tried to remove the flaw by being more careful with the logic or the philosophy. Gödel's proof is essentially a formal version of one proposed by Leibniz. Gödel never published his proof because he was worried that it might be seen as a rigorous demonstration of the existence of God, whereas he viewed it as a formal statement of Leibniz's tacit assumptions, which would help to reveal potential logical errors. For further analysis see en.wikipedia.org/wiki/G%C3%B6del's_ontological_proof and for a detailed discussion of modal logic and its use in the proof see www.stats.uwaterloo.ca/~cgsmall/ontology1.html

Answers on page 310

Professor Stewart's Cunning Crib Sheet

Wherein the discerning or desperate reader may
locate answers to those questions that are
currently known to possess them ... with
occasional supplementary facts for
their further edification.

Alien Encounter

Alfy is a Veracitor, whereas Betty and Gemma are Gibberish.

There are only eight possibilities, so you can try each in turn. But there's a quicker way. Betty said that Alfy and Gemma belong to the same species, but they have given different answers to the same question, so Betty is Gibberish. Alfy said precisely that, making him a Veracitor. Gemma said the opposite, so she must be Gibberish.

Curious Calculations

$$1 \times 1 = 1$$
$$11 \times 11 = 121$$
$$111 \times 111 = 12,321$$
$$1,111 \times 1,111 = 1,234,321$$
$$11,111 \times 11,111 = 123,454,321$$

If you know how to do 'long multiplication', you can see why this striking pattern occurs. For instance,

$$111 \times 111 =$$
$$11,100 +$$
$$1,110 +$$
$$111$$

We find one '1' in the units column, two in the tens column, three in the hundreds; then the numbers shrink again, with two in the thousands and one in the ten thousands. So the answer must be 12,321.

The pattern does continue – but your calculator may run out of digits. In fact,

$$111,111 \times 111,111 = 12,345,654,321$$
$$1,111,111 \times 1,111,111 = 1,234,567,654,321$$
$$11,111,111 \times 11,111,111 = 123,456,787,654,321$$
$$111,111,111 \times 111,111,111 = 12,345,678,987,654,321$$

After this the pattern breaks down, because digits 'carry' and spoil it.

$$142,857 \times 2 = 285,714$$
$$142,857 \times 3 = 428,571$$
$$142,857 \times 4 = 571,428$$
$$142,857 \times 5 = 714,285$$
$$142,857 \times 6 = 857,142$$
$$142,857 \times 7 = 999,999$$

When we multiply 142,857 by 2, 3, 4, 5 or 6, we get the same sequence of digits in cyclic order, but starting at a different place. The 999,999 is a bonus.

This curious fact is not an accident. Basically, it happens because 1/7 in decimals is 0.142 857 142 857 . . ., repeating for ever.

Triangle of Cards

The 15-card difference triangle.

Turnip for the Books

Hogswill started with 400 turnips.

The way to solve this kind of puzzle is to work backwards.

Suppose that at the start of hour 4, Hogswill has x turnips. By the end of the hour he has sold $\frac{6x}{7} + \frac{1}{7}$ turnips, and since none are left, this equals x. So $x - \frac{6x}{7} + \frac{1}{7} = \frac{x-1}{7} = 0$, and $x = 1$. Similarly, if he had x turnips at the start of hour 3, then $\frac{x-1}{7} = 1$, so $x = 8$. If he had x turnips at the start of hour 2, then $\frac{x-1}{7} = 8$, so $x = 57$. Finally, if he had x turnips at the start of hour 1, then $\frac{x-1}{7} = 57$, so $x = 400$.

The Four-Colour Theorem

Here are the four counties for which each is adjacent to all of the others. The middle one is West Midlands – which coincidentally is where I live – and the three surrounding it are Staffordshire, Warwickshire and Worcestershire, clockwise from the top.

These counties imply that we need at least four colours.

Shaggy Dog Story

First, the dodgy arithmetic. The method 'works' because the will's terms are inconsistent. The fractions do not add up to 1. In fact,

$$\frac{1}{2} + \frac{1}{3} + \frac{1}{9} = \frac{17}{18}$$

which should make the trick obvious.

Whoever first designed this puzzle was clever – there are very few numbers that work, and this choice disguises the inconsistency very neatly. I mean, how would you feel about a puzzle where the uncle has 1,129 dogs, the sons are bequeathed $\frac{4}{7}$, $\frac{3}{11}$ and $\frac{2}{15}$ of them, and Lunchalot rides to the rescue with 26 extra dogs?

However, there is another neat possibility: exactly the same, except that the third son gets one-seventh of the dogs. If the same trick works, how many dogs were there?

Answer to the Answer

The clue is that

$$\frac{1}{2} + \frac{1}{3} + \frac{1}{7} = \frac{41}{42}$$

So there were 41 dogs.

Answer Continued

Oops, I nearly forgot the actual *question*: what did Gingerbere say to Ethelfred that so offended Sir Lunchalot?

It was this: 'Surely you wouldn't send a knight out on a dog like this?'

I said it was a shaggy dog story.

Confession

The shaggy dog story is inspired, in part, by the science fiction short story 'Fall of knight' by A. Bertram Chandler, which appeared in *Fantastic Universe* magazine in 1958.

Rabbits in the Hat

Nothing is wrong with the calculation, but its interpretation is nonsense. When the various probabilities are combined, we are working out the probability of extracting a black rabbit, over *all possible combinations* of rabbits. It is fallacious to imagine that this probability is valid for any specific combination. The fallacy is glaring if there is only one rabbit in the hat. With one rabbit, a similar argument (ignoring adding and removing a black rabbit, which changes nothing essential) goes like this: the hat contains either B or W, each with probability $\frac{1}{2}$. The probability of extracting a black rabbit is therefore

$$\frac{1}{2} \times 1 + \frac{1}{2} \times 0$$

which is $\frac{1}{2}$. Therefore (*really?*) half the rabbits in the hat are black, and half are white.

But there's only one rabbit in the hat ...

River Crossing 1 – Farm Produce

There are two solutions. One is:

(1) Take the goat across.

(2) Come back with no cargo, pick up the wolf, and take that across.

(3) Bring the goat back, but leave the wolf.

(4) Drop off the goat, pick up the cabbage, cross the river, leave the cabbage.

(5) Come back with no cargo, pick up the goat, take it across.

In the other, the roles of wolf and cabbage are exchanged.

I like to solve this geometrically, using a picture in *wolf–goat–cabbage space*. This consists of triples (w, g, c) where each symbol is either 0 (on this side of the river) or 1 (on the far side). So, for instance, $(1, 0, 1)$ means that the wolf and cabbage are on the far side but the goat is on this side. The problem is to get from $(0, 0, 0)$ to $(1, 1, 1)$ without anything being eaten. We don't need to say where the farmer is, since he always travels in the boat during river crossings.

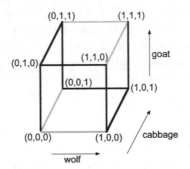

Wolf–goat–cabbage space: now it's obvious.

There are eight possible triples, and they can be thought of as the vertices of a cube. Because only one item can accompany the farmer on each trip, the permissible moves are the edges of the cube.

However, four edges (shown in grey) are not permitted, because things get eaten. The remaining edges (black) do not cause mayhem.

So the puzzle reduces to a geometric one: find a route along the black edges, from $(0, 0, 0)$ to $(1, 1, 1)$. The two solutions are immediately evident.

More Curious Calculations

(1) $13 \times 11 \times 7 = 1,001$, and this is why the trick works. If you multiply a three-digit number *abc* by 1,001 the result is *abcabc*. Why? Well, multiplying by 1,000 gives *abc*000. Then you add a final *abc* to multiply by 1,001.

(2) For four-digit numbers, everything is similar, but we have to multiply by 10,001. This can be done in two stages – multiply by 73 and then by 137 – because $73 \times 137 = 10,001$.

(3) For five-digit numbers we have to multiply by 100,001. This can be done in two stages – multiply by 11 and then by 9,091 – because $11 \times 9,091 = 100,001$. As a party trick, this is a bit contrived, though.

(4) We get 471,471,471,471 – the same three digits repeated four times. Why? Because

$$7 \times 11 \times 13 \times 101 \times 9,901 = 1,001,001,001$$

(5) Adding the final 128 leads to 128,000,000 – a million times the original number. This trick works for all three-digit numbers, and it does so because

$$3 \times 3 \times 3 \times 7 \times 11 \times 13 \times 37 = 999,999$$

Add 1 and you get a million.

You can turn all these tricks into party magic tricks. For instance, the trick that turns 471,471 into 471 could be presented like this. The magician, with eyes blindfolded, asks a member of the audience to write down a three-digit number (say 471) on a blackboard or a sheet of paper. A second person then writes it down twice (471,471). A third, armed with a calculator, divides that by 13 (getting 36,267). A fourth divides the result by 11 (getting 3,297). While this is going on, the magician makes a lot of fuss about how unlikely it is that either of these numbers divides without remainder. Then she asks what the final result is, and instantly announces that the original number was 471.

To work this out, she mentally divides 3,297 by 7. OK, you have to be able to do that, but if you know your seven times table it's easy.

Extracting the Cherry

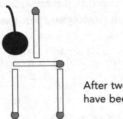

After two matches
have been moved.

Make Me a Pentagon

Tie a knot in the strip, and flatten it – carefully.

Pentagon from
knotted strip.

An interesting challenge is to prove that the result really is a regular pentagon – in an idealised Euclidean version of the problem. I'll leave that for anyone who is interested.

Empty Glasses

Pick up the second glass from the left, pour its contents into the fifth glass, and replace the second glass.

Three Quickies

(1) If you and your partner hold all the spades, your opponents

hold none, and vice versa. So the likelihood is the same in each case.

(2) Three. You *took* three, so that's how many you have.

(3) Zero. If five are in the right envelope, so is the sixth.

Knight's Tours

There is no closed tour on the 5×5 square. Imagine colouring the squares black and white in the usual chessboard fashion. Then the knight changes colour at each move. A closed tour must then have equal numbers of black and white squares. But $5 \times 5 = 25$ is odd. The same argument rules out closed tours on all squares with odd sides.

There is no tour on the 4×4 square. The main obstacle is that each corner square connects to only two other squares, and the diagonally opposite corner *also* connects to those two squares. A little thought proves that if a tour of all 16 squares exists, it must start at one corner and finish in an adjacent corner. Systematic consideration of possibilities shows that this is impossible.

However, there is a tour that visits 15 of the 16 squares (showing that the situation regarding a complete tour is delicate):

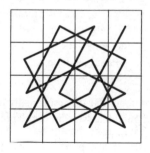

How the knight can visit 15 squares.

White-Tailed Cats

Suppose that there are c cats, of which w have white tails. There are c $(c-1)$ ordered pairs of distinct cats, and $w(w-1)$ ordered pairs of white-tailed cats. (You can choose the first cat of the pair in c ways, but the second in only $c-1$ ways since you've used up one cat. Ditto for white-tailed cats. By 'ordered' I mean that choosing first cat A and then cat B is considered to be different from B first and then A. If you don't like that, then both formulas have to be halved – with the same result.)

This means that the probability of *both* cats having white tails is

$$\frac{w(w-1)}{c(c-1)}$$

and this must be $\frac{1}{2}$. Therefore

$$c(c-1) = 2w(w-1)$$

with c and w whole numbers. The smallest solution is $c = 4$, $w = 3$. The next smallest turns out to be $c = 21$, $w = 15$. Since Ms Smith has fewer than 20 cats, she must have four cats, of which three have white tails.

Perpetual Calendar

Each cube must include 1 and 2 so that 11 and 22 can be represented. If only one cube bears a 0, then at most six of the nine numbers 01–09 can be represented, so both must bear a 0 as well. That leaves six spare faces for the seven digits 3–9, so the puzzle looks impossible ... until you realise that the cube bearing the number 6 can be turned upside down to represent 9. So the white cube bears the numbers 0, 1, 2, 6 (also 9), 7 and 8, and the grey cube bears the numbers 0, 1, 2, 3, 4 and 5. (Note that I've shown a 5 on my grey cube, and that tells us which cube is which.)

Deceptive Dice

There is no best dice. If Innumeratus plays, and Mathophila chooses correctly (as she will, because she's like that), then he will lose, in the long run. The odds will always favour Mathophila.

How come? Mathophila has constructed her dice so that on average, the yellow one beats the red one, the red one beats the blue one – and the blue one beats the yellow one! At first sight this seems impossible, so let me explain why it's true.

Each number occurs twice on each of the dice, so the chance of rolling any particular number is always $\frac{1}{3}$. So I can make a table of the possibilities, and see who wins for which combinations of numbers thrown. Each combination has the same probability, $\frac{1}{9}$.

Yellow versus red

	1	5	9
3	Red	Yellow	Yellow
4	Red	Yellow	Yellow
8	Red	Red	Yellow

Here yellow wins five times out of nine, red wins only four times.

Red versus blue

	3	4	8
2	Red	Red	Red
6	Blue	Blue	Red
7	Blue	Blue	Red

Here red wins five times out of nine, blue wins only four times.

Blue versus yellow

	2	6	7
1	Blue	Blue	Blue
5	Yellow	Blue	Blue
9	Yellow	Yellow	Yellow

Here blue wins five times out of nine, yellow wins only four times.

So yellow beats red $\frac{5}{9}$ of the time, red beats blue $\frac{5}{9}$ of the time, and blue beats yellow $\frac{5}{9}$ of the time.

This gives Mathophila an advantage if she chooses *second*, which she has cunningly arranged. If Innumeratus chooses the red dice, she should choose yellow. If he chooses the yellow dice, she should choose blue. And if he chooses the blue dice, she should choose red.

It may not be a huge advantage – five chances out of nine of winning, compared with four out of nine – but it's still an advantage.

262 // Professor Stewart's Cunning Crib Sheet

In the long run, Innumeratus will lose his pocket money. If he wants to play, then a gentlemanly 'No, *you* choose' would be a good idea.

It may seem impossible to have yellow 'better than' red, and red 'better than' blue – but not to have yellow 'better than' blue. What's happening is that the meaning of 'better than' depends on which dice are being used. It's a bit like three football teams:

- The Reds have a good goalkeeper and a good defence, but a poor attack. They win if and only if the opposing goalie is poor.
- The Yellows have a poor goalie, a good defence, and a good attack. They win if and only if the opposing defence is poor.
- The Blues have a good goalie, a poor defence, and a good attack. They win if and only if the opposing attack is poor.

Then (check this!) the Reds always beat the Yellows, the Yellows always beat the Blues, and the Blues always beat the Reds.

Dice like this are said to be *intransitive*. ('Transitive' means that if A beats B and B beats C then A beats C. That doesn't happen here.) On the practical side, the existence of intransitive dice tells us that some 'obvious' assumptions about economic behaviour are actually wrong.

An Age-Old Old-Age Problem

Scrumptius was 69. There was no year 0 between BC dates and AD dates. (If you decided that he might be 68 if he died earlier in the day than he was born, you get a point for ingenuity. But you lose two for pedantry, because it is usual to increase a person's age by one year as soon as their birthday begins, immediately after midnight.)

Heron Suit

The deduction is incorrect. Consider a cat with blunt claws that plays with a gorilla, does not wear a heron suit, has a tail, has no whiskers and is unsociable. The first five statements are all true, but the sixth is not.

I'd explain about the heron suit, but my cat has refused permission on the grounds that it might incriminate itself.

How to Unmake a Greek Cross

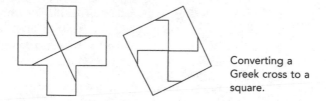

Converting a
Greek cross to a
square.

Euler's Pentagonal Holiday

Here's a solution to (b), which is automatically a solution to (a) as
well. There are others. But they all have to start and end at the two
vertices with valency 3, and a mirror-symmetric one must always
have the bottom edge of the pentagon in the middle of the tour.

A solution with
left-right symmetry.

Ouroborean Rings

One possible ouroborean ring for quadruplets is

   1111000010100110

There are others. The topic has a long history, going back to Irving
Good in 1946. Ouroborean rings exist for all m-tuples of n digits: for
example, in this one

   000111222121102202101201002

each triple of the three digits 0, 1, 2 occurs exactly once.

   How many ouroborean rings are there? In 1946 Nicholas de
Bruijn proved that for m-tuples formed from the two digits 0 and 1,
this number is $2^{2^{m-1}-m}$, which grows extremely fast. Here rings
obtained by rotating a given one are considered to be the same.

m	Number of ouroborean rings
2	1
3	2
4	16
5	2,048
6	67,108,864
7	144,115,188,075,855,872

The Ourotorus

There is a unique solution, except for various symmetry transformations – rotation, reflection and translations horizontally or vertically. Bear in mind the 'wrap round' convention. So you can, for instance, cut off the four pieces on the right and move them to the left.

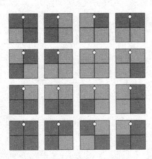

Solution to the ourotorus puzzle.

A Constant Bore

The only reason for including this kind of question in this kind of book is if something surprising happens, and the only surprising thing that makes much sense is that the answer does *not* depend on the radius of the sphere.

That sounds crazy – suppose the sphere were the Earth? But to make the hole only 1 metre long, you have to remove almost the entire planet, leaving only a *very* thin band round the equator, one metre wide. So just maybe . . .

Here comes the easy bit. Assuming that the radius really does not matter, we can work out the answer by considering the special case when the hole is very narrow – in fact, when its width is zero.

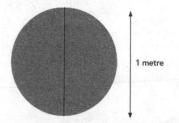

1 metre

Special case of the problem.

Now the volume of copper is equal to that of the entire sphere, and the diameter of the sphere is 1 metre. So its radius is $r = \frac{1}{2}$, and its volume is given by the famous formula

$$V = \frac{4}{3}\pi r^3$$

which equals $\pi/6$ when $r = \frac{1}{2}$.

Ah, but how do we *know* that the answer doesn't depend on the radius? That's a bit more complicated, and it uses more geometry. (Or you can do it by calculus, if you know how.)

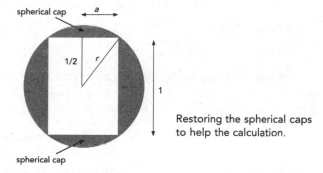

spherical cap

a

1/2 r

1

spherical cap

Restoring the spherical caps to help the calculation.

Put back the missing 'spherical caps' on the top and bottom. Suppose that the radius of the sphere is r, and the radius of the cylindrical hole is a. Then Pythagoras's Theorem applied to the small triangle at the top right tells us that

$$r^2 = a^2 + \left(\tfrac{1}{2}\right)^2$$

so

$$a^2 = r^2 - \tfrac{1}{4}.$$

Now we need three volume formulas:

- The volume of a sphere of radius r is $\tfrac{4}{3}\pi r^3$.
- The volume of a cylinder of base radius a and height h is $\pi a^2 h$.
- The volume of a spherical cap of height k in a sphere of radius r is $\tfrac{1}{3}\pi k^2(3r - k)$.

Don't worry, I had to look that last one up myself.

The volume of copper required is the volume of the sphere, minus that of the cylinder, minus that of two spherical caps, which is

$$\tfrac{4}{3}\pi r^3 - \pi a^2 h - \tfrac{2}{3}\pi k^2(3r - k)$$

since there are two spherical caps. But $h = 1, k = r - \tfrac{1}{2}$ and $a^2 = r^2 - \tfrac{1}{4}$, so the volume is

$$\tfrac{4}{3}\pi r^3 - \pi\left(r^2 - \tfrac{1}{4}\right) - \tfrac{2}{3}\pi\left(r - \tfrac{1}{2}\right)^2\left[3r - \left(r - \tfrac{1}{2}\right)\right] = \left[2r + \tfrac{1}{2}\right]$$

Doing the algebra, almost everything miraculously cancels, and all that remains is $\pi/6$.

Digital Century

$$123 - 45 - 67 + 89 = 100$$

This solution was found by the great English puzzle-creator, Henry Ernest Dudeney, and can be found in his book *Amusements in Mathematics*. There are lots of answers if you use four or more arithmetical symbols.

Squaring the Square

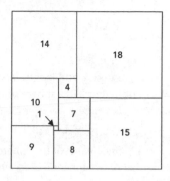

How Morón's tiles make a rectangle.

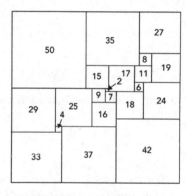

How Duijvestijn's tiles make a square.

You can also rotate or reflect these arrangements.

Ring a-Ring a-Ringroad

The difference is 20π metres, or roughly 63 metres, for roads on the flat. It doesn't depend on the length of the motorway, or how wiggly it is, provided the curvature is gradual enough for '10 metres distance between lanes' to be unambiguous.

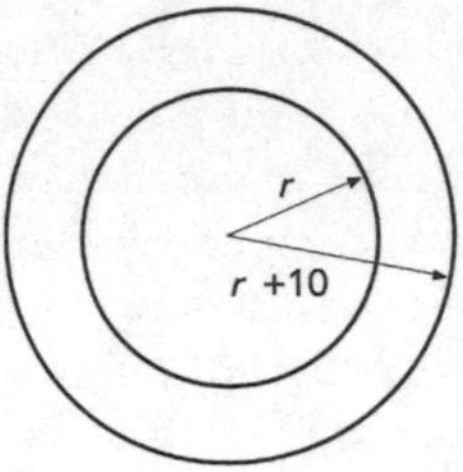

Data for a circular M25.

Let's start with an idealised version, where the M25 is a perfect circle. If the anticlockwise lane has radius r, then the clockwise one has radius $r + 10$. Their circumferences are then $2\pi r$ and $2\pi(r + 10)$. The difference is

$$2\pi(r + 10) - 2\pi r = 20\pi$$

which is independent of r.

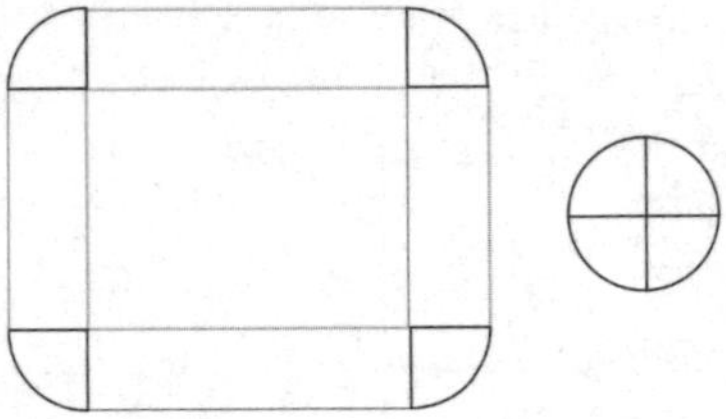

A rectangular motorway also creates an excess of 20.

However, the M25 is not circular. For argument's sake, try a rectangle. Now the outer lane consists of four straight bits, which match the inner lane exactly, plus four quarter-circles at the corners. These extra arcs fit together to make a single circle of radius 10. Again, we get an excess of exactly 20π.

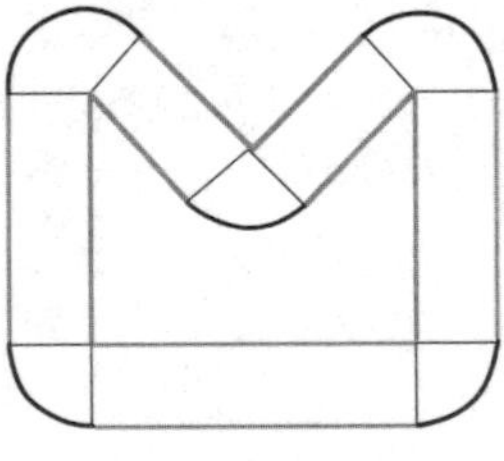

A non-convex polygon gives 20π as well.

The same point holds for any 'polygonal' road – one composed of straight lines, plus arcs of circles at corners. The straight-line parts match; the arcs add up to one complete circle of radius 10. This is true even when the polygon is not convex, such as the M-shape shown above.* Now the outer lane has arcs that add up to one and a quarter circles, and the inner lane has a quarter circle of its own. But this quarter-turn is of opposite curvature, so it cancels out the excess quarter-turn in the outer lane. The point is that any sufficiently smooth curve can be approximated as closely as we wish by polygons, so the excess is 20π in *all* cases.

The same argument applies to runners on a curved track. In the 400 metres, runners start from 'staggered' positions, to make the overall distance the same in each lane. The stagger between adjacent lanes must be 2π times the width of a lane. This width is usually 1.22 metres, so the stagger should be 7.66 metres per lane – provided it is applied on a straight section of the track. In practice the region where the athletes start often includes part of a bend, so the numbers are a bit different. The easy way to calculate them is to make sure that each runner goes exactly
400 metres, which is what the rules actually state.

Magic Hexagon

The only solution (apart from rotations and reflections of it) is

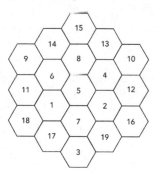

The only non-trivial magic hexagon.

This magic hexagon was found independently by several people

---

* It's not true if the polygon crosses itself, as does the Suzuka racing circuit in Japan. But for some reason figure-of-eight orbital motorways don't seem to have caught on.

between 1887 and 1958. If we try similar patterns of hexagons with n cells along the edge instead of 3, then the only other case where a magic hexagon (using consecutive numbers 1, ..., n) exists is the trivial pattern when $n = 1$: a single hexagon containing the number 1. Charles W. Trigg explained why in 1964, by proving that the magic constant must be

$$\frac{9(n^4 - 2n^3 + 2n^2 - n) + 2}{2(2n - 1)}$$

which is an integer only when $n = 1$ or 3.

Pentalpha

The star shape is designed to mislead. The important aspect of the structure is which circles are two steps away from which, because these are where each new counter starts and finishes. By focusing on this we can draw a much simpler diagram:

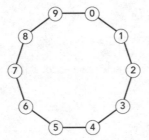

A transformed version of the puzzle.

The rule for placing counters is now: place each new counter on an empty circle and slide it to an adjacent empty circle. It is now obvious how to cover nine circles. For example, place a counter on 1 and slide it to 0. Then place a counter on 2 and slide it to 1. Then place a counter on 3 and slide it to 2. Continue in this way, placing each new counter two empty dots away from the existing string of counters.

Copy these moves on the original diagram to solve the puzzle.

In the second diagram, you can add new counters at either end, so there are lots of solutions. But you can't create more than one connected chain of counters at any stage, because there are then at least two gaps where no counters exist, and each gap leads to at least one circle that can't be covered.

How Old Was Diophantus?

Diophantus was 84 when he died. Let x be his age. Then

$$\frac{x}{6} + \frac{x}{12} + \frac{x}{7} + 5 + \frac{x}{2} + 4 = x$$

So

$$\frac{9}{84}x = 9$$

and $x = 84$.

The Sphinx is a Reptile

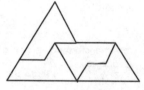

Four sphinxes make a bigger sphinx.

Langford's Cubes

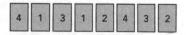

Langford's cubes with four colours.

Magic Stars

This arrangement – possibly rotated or reflected – is the only solution.

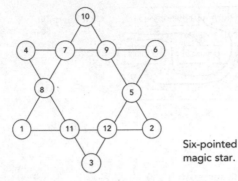

Six-pointed
magic star.

Curves of Constant Width

Surprisingly, the circle isn't the only curve of constant width. The simplest curve of constant width that is not a circle is an equilateral triangle with rounded edges:

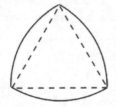

(Left) Constant-width triangle.

(Right) Twenty-pence coin.

Each edge is an arc of a circle, with centre at the opposite vertex. Two British coins, the 20p and 50p, are 7-sided curves of constant width; this shape was chosen because it makes the coins suitable for use in slot machines, but distinguishes them from other circular coins worth different amounts – which is especially useful for the visually impaired.

Connecting Cables

The main point is *not* to connect the dishwasher first, with a straight cable. This isolates each of the other appliances from its socket, and makes a solution impossible. If you connect the fridge and cooker first, it's then obvious how to hook up the dishwasher.

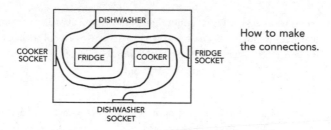

How to make the connections.

Coin Swap

One solution is to successively swap the following pairs: HK, HE, HC, HA, IL, IF, ID, KL, GJ, JA, FK, LE, DK, EF, ED, EB, BK. There are many others.

The Stolen Car

Fenderbender paid out £900 for the car and an extra £100 to the clergyman as change. He counted all his outgoings but forgot to include the corresponding income. All other transactions cancel out, so he lost £1,000.

Compensating Errors

The numbers were 1, 2 and 3. Then $1 + 2 + 3 = 6 = 1 \times 2 \times 3$. This is the only solution for three positive whole numbers.

With two numbers, the only possibility is $2 + 2 = 4 = 2 \times 2$. With four numbers, the only solution is $1 + 1 + 2 + 4 = 8 = 1 \times 1 \times 2 \times 4$.

With more numbers, there are usually lots of solutions, but in some exceptional cases there is just one solution. If the sum of k positive whole numbers is equal to their product, and only *one* set of k numbers has that property, then k is one of the numbers 2, 3, 4, 6, 24, 114, 174 and 444, or it is at least 13, 587, 782, 064. No examples greater than that are known, but their possible existence remains open.

River Crossing 2 – Marital Mistrust

A graphical solution is a bit messy to draw because it involves a 6-dimensional hypercube in husband$_1$–husband$_2$–husband$_3$–wife$_1$–

wife$_2$–wife$_3$ space. Fortunately there's an alternative. Eliminating unsuitable moves and using a bit of logic leads to a solution in 11 moves, which is the smallest possible number. Here the husbands are *A B C* and the corresponding wives are *a b c*.

This bank	In boat	Direction	Far bank
A C a c	*B b*	→	—
A C a c	*B*	←	*b*
A B C	*a c*	→	*b*
A B C	*a*	←	*b c*
A a	*B C*	→	*b c*
A a	*B b*	←	*C c*
a b	*A B*	→	*C c*
a b	*c*	←	*A B C*
b	*a c*	→	*A B C*
b	*B*	←	*A C a c*
—	*B b*	→	*A C a c*

There are minor variations on this solution in which various couples are interchanged.

Wherefore Art Thou Borromeo?

In the second pattern, the two lower rings are linked. In the third pattern, all three pairs are linked. In the fourth pattern, the top ring is linked to the left one which in turn is linked to the right one.

There are lots of four-ring versions. Here's one:

A set of four
'Borromean' rings.

Analogous arrangements exist for any finite number of rings. It has been proved that the Borromean property can't be obtained using perfectly circular (and therefore flat) rings. This is a topological phenomenon.

Percentage Play

The profit and loss do not cancel out. The bicycle he sold to Bettany cost him £400 (he lost £100, which is 25% of £400). The one he sold to Gemma cost him £240 (he gained £60, which is 25% of £240). Overall, he paid £640 and received £600, so he lost £40.

New Merology

Assign the values

E	F	G	H	I	L	N	O	R	S	T	U	V	W	X	Z
3	9	6	1	−4	0	5	−7	−6	−1	2	8	−3	7	11	10

Then

$$Z + E + R + O = 0$$
$$O + N + E = 1$$
$$T + W + O = 2$$
$$T + H + R + E + E = 3$$
$$F + O + U + R = 4$$
$$F + I + V + E = 5$$
$$S + I + X = 6$$
$$S + E + V + E + N = 7$$
$$E + I + G + H + T = 8$$

$$N + I + N + E = 9$$
$$T + E + N = 10$$
$$E + L + E + V + E + N = 11$$
$$T + W + E + L + V + E = 12$$

Spelling Mistakes

There are four *spelling* mistakes, in the words 'there', 'mistakes', 'in' and 'sentence'. The fifth mistake is the claim that there are five mistakes, when there are really only four.

But ... this means that if the sentence is true, it has to be false, but if it's false, it has to be true. Oops.

Expanding Universe

Perhaps surprisingly, the *Indefensible* actually *does* get to the edge of the universe ... but it takes about 10^{434} years to do so. By then the universe has grown to a radius of about 10^{437} light years.

Let's see why.

At each stage, when the universe expands, the *fraction* of the distance that the *Indefensible* has already covered doesn't change. That suggests that if we think about the fractions, we ought to be able to find the answer more easily.

In the first year the ship travels 1/1,000 of the distance to the edge. In the next year it travels 1/2,000 of the distance. In the third year it travels 1/3,000 of the distance, and so on. In the nth year it travels $1/1,000n$ of the distance. So the total fraction travelled after n years is

$$\frac{1}{1,000} \left(1 + \frac{1}{2} + \frac{1}{3} + \frac{1}{4} + \ldots + \frac{1}{n} \right) = \frac{1}{1,000} H_n$$

which is why harmonic numbers are relevant. In particular, the number of years required to reach the edge is whatever value of n first makes this fraction bigger than 1 – that is, makes H_n bigger than 1,000. There is no known formula for the value of H_n in terms of n, and it grows very slowly as n increases. However, it can be proved that by making n large enough, H_n can be made as large as we wish – and in particular, greater than 1. So the *Indefensible* does get to the edge if n is sufficiently large.

To find out how large, we use the hint. To make $H_n > 1,000$ we require $\log n + \gamma > 1,000$, so that $n > e^{1,000-\gamma}$. So the number of years required to reach the edge of the universe is very close to $e^{999.423}$, which is 10^{434} in round numbers. By then the universe will have grown to $n + 1$ thousand light years, which is near enough 10^{437} light years.

Initially the remaining distance keeps increasing each year, but eventually the ship starts to catch up with the expanding edge of the universe. Its 'share' of the expansion grows as it gets farther out, and in the long run this beats the fixed expansion rate of 1,000 light years per year of the edge of the universe. The 'long run' here is very long: it takes about $e^{999-\gamma} = 10^{433.61}$ years before the remaining distance starts to decrease – roughly the first third of the voyage.

Family Occasion

The smallest possible number of party guests is seven: two small girls and one boy, their father and mother, and their father's father and mother.

Don't Let Go!

Your body plus the rope forms a closed loop. It is a topological theorem that a knot cannot be created in a closed loop by deforming it continuously, so the problem can never be solved if you pick up the rope in the obvious 'normal' way. Instead, you must first tie a knot *in your arms*. This may sound difficult, but anyone can do it: just fold them across your chest. Now lean forward so that the hand that is on top of an arm can reach over the arm to pick up one end of the rope, and pick up the other end with the other hand. Unfold your arms, and the knot appears.

Möbius and His Band

If you cut a Möbius band along the middle, it stays in one piece – see the second limerick. The resulting band has a 360° twist.

If you cut a Möbius band one-third of the way across, you get two linked bands. One is a Möbius band, the other (longer) one has a 360° twist.

If you cut a band with a 360° twist along the middle, you get two linked bands with 360° twists.

Three More Quickies

(1) Five days. (Each dog digs a hole in five days.)

(2) The parrot is deaf.

(3) The usual answer is that one hemisphere of the planet is land, and the other is water, so the continent and the island are

identical. But puzzles like this are easily 'cooked' by finding loopholes in the conditions. For instance, maybe Nff lives on the continent but its house is on the island, and Pff eats houses for breakfast. Or on Nff-Pff, the land moves – after all, who knows what happens on an alien world? *Or* ...

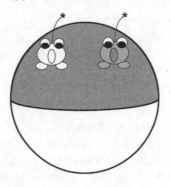

Nff and Pff on their home planet of Nff-Pff.

Miles of Tiles

I forgot to add an extra condition: the tiles should meet at their corners. The corners of some might meet the edges of others. This doesn't change the answer, but it complicates the proof a little.

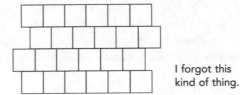

I forgot this kind of thing.

Après-le-Ski

The cables cross at a height of 240 metres.

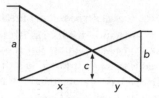

More generally...

It's simpler to tackle a more general problem, where the lengths are as shown. By similar triangles,

$$\frac{x+y}{a} = \frac{y}{c} \quad \text{and} \quad \frac{x+y}{b} = \frac{x}{c}$$

Adding, we get

$$(x+y)\left(\frac{1}{a}+\frac{1}{b}\right) = \frac{x+y}{c}$$

Dividing by $x+y$, we obtain

$$\frac{1}{a} + \frac{1}{b} = \frac{1}{c}$$

leading to

$$c = \frac{ab}{a+b}$$

We notice that c does not depend on x or y, which is a good job since the puzzle didn't tell us those. We know that $a = 600$, and $b = 400$, so

$$c = \frac{600 \times 400}{1000} = 240$$

Pick's Theorem

The lattice polygon illustrated has $B = 21$ and $I = 5$, so its area is $14\frac{1}{2}$ square units.

Paradox Lost

I don't think this one stands up to scrutiny. Both litigants are doing a pick-and-mix – at one moment assuming that the agreement is valid, but at another, assuming that the court's decision can override the agreement. But why do you take an issue like this to court? Because the court's job is to resolve any claimed ambiguities in the contract, *override the contract if need be*, and tell you what to do. So if the court orders the student to pay up, then he has to, and if the court says that he doesn't have to pay up, then Protagoras doesn't have a leg to stand on.

Six Pens

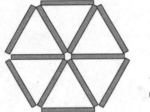

12 panels
making 6 pens.

Hippopotamian Logic

Therefore oak trees grow in Africa.

Why? Suppose, on the contrary, that oak trees *don't* grow in Africa. Then squirrels hibernate in the winter, and hippos eat acorns. Therefore I'll eat my hat. But I won't eat my hat, a contradiction. Therefore (*reductio ad absurdum*) my assumption that oak trees don't grow in Africa must be false. So oak trees grow in Africa.

Pig on a Rope

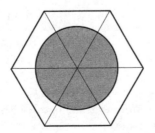

Six copies simplify
the geometry.

To simplify the problem, make six copies of the field, with six copies of the (shaded) region accessible to the pig. Then we want the shaded circle to have half the area of the hexagon. The area of the circle is πr^2 where r is the radius. The hexagon has sides 100 metres long, so its area is $10,000 \times 3\sqrt{3}/2$, or $15,000\sqrt{3}$. So $\pi r^2 = 7,500\sqrt{3}$, and

$$r = \sqrt{\left(\frac{7,500\sqrt{3}}{\pi}\right)}$$

which is about 64.3037 metres.

The Surprise Examination

I think that the Surprise Examination Paradox is a very interesting case of something that looks like a paradox but isn't. My reason is that there is a logically equivalent statement, which is obviously true – but totally uninteresting.

Suppose that every morning the students announce confidently, 'The test will be today.' Then eventually they will do so on the day of the actual test, at which point they will be able to claim that the test was not a surprise.

I don't see any logical objection to this technique, but it's obviously a cheat. If you expect something to happen every day, then of course you won't be surprised when it does. My view – and I've argued with enough mathematicians who didn't agree with me, let alone anybody else, so I'm aware that there's room for differences of opinion – is that the paradox is bogus. It is nothing more than this obvious strategy, dressed up to look mysterious. The cheat is not entirely obvious, because everything is intuited instead of being acted upon, but it's the *same* cheat.

Let me sharpen the conditions by requiring the students to state, each morning before school begins, whether they think the test will be held that day. With this condition, in order for the students to *know* that it can't be on Friday, they have to leave themselves the option of announcing on Friday morning that 'It will be today.' And the same goes for Thursday, Wednesday, Tuesday and Monday. So they have to say 'It will be today' five times in all – once per day. This makes sense: if the students are allowed to revise their prediction each day, then eventually they'll be right.

If we demand even the tiniest bit more, though, their strategy falls to bits. For example, suppose that they're allowed only one such announcement. If Friday arrives and they haven't used up their guess, then they can make the announcement then. But if they *have* used up their guess, they're in trouble. Worse, they *can't* wait until Friday to use their guess, because the test might be on Monday, Tuesday, Wednesday or Thursday.

In fact, if they are allowed *four* guesses, they're still sunk. Only if they are permitted five guesses can they guarantee to predict the correct day. But any fool could do that.

I'm proposing two things here. The less interesting one is that the paradox hinges on what we mean by 'surprise'. The more interesting proposal is that *whatever* we mean by 'surprise', there are two logically equivalent ways to state the students' prediction strategy. One – the usual presentation – seems to indicate a genuine paradox. The other – describe the strategy in terms of actual actions, not hypothetical ones – turns it into something correct but unsurprising, destroying the element of paradox.

Equivalently, we can up the ante by letting the teacher add another condition. Suppose that the students have poor memories, so that any work they do on a given evening to prepare for the test is forgotten by the next evening. If, as the students claim, the test is not going to be a surprise, then they ought to be able to get away with very little homework: just wait until the evening before the test, then cram, pass and forget. But the teacher, in her wisdom, knows that this won't work. If they don't do their homework on Sunday evening, the test could be on Monday, and if it is, they'll fail. Ditto Tuesday through Friday. So despite claiming never to be surprised by the test, the students have to do five evenings of homework.

Antigravity Cone

The uphill motion is an illusion. As the cone moves in the 'uphill' direction, its centre of gravity moves downwards, because the slope widens out and the cone is supported nearer to its two ends.

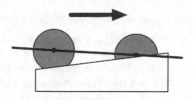

Side view: as the cone follows the arrow, which points *up* the slope, its centre of gravity moves *down* (black line).

What Shape is a Crescent Moon?

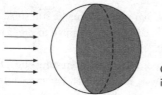

Geometry of an illuminated sphere.

The left curve of the crescent is a semicircle, but the other edge is not an arc of a circle. It is a 'semiellipse' – an ellipse cut in half along its longest axis. The diagram shows the rays of the Sun, which we are assuming to be parallel. In this view the Sun must be positioned some distance *behind* the plane of the page to create the crescent. The light and dark portions of the Moon are *hemispheres*, so the boundary between them is a circle; in fact, it is where a plane at right angles to the Sun's rays cuts the sphere. We observe this circle at an angle. A circle viewed at an angle is an ellipse – its hidden edge is drawn dotted, and we see only the front half. (I've used grey shading to show the dark part of the moon.)

In reality the illumination becomes very faint near the boundary between light and dark, and the Moon is a bit bumpy. So the shape is not as clearly defined as this discussion suggests. You can also quibble about how the circle is projected on to the retina if you so desire.

The crescent shape formed by two circular arcs *can* sometimes be seen in the sky – most dramatically during an eclipse of the Sun, when the Moon partially overlaps the Sun's disk. But now it is the Sun, not the Moon, that looks like a crescent.

Famous Mathematicians

The odd famous mathematician out is Carol Vorderman – see below.

Pierre Boulez Modernist composer and conductor. Studied mathematics at the University of Lyon but then switched to music.

Sergey Brin Co-founder of Google™, with Larry Page. Computer Science and Mathematics degree from University of Maryland. Net worth estimated at $16.6 billion in 2007, making him the 26th-richest person in the world. The Google search engine is based on mathematical principles.

Lewis Carroll Pseudonym of Charles Lutwidge Dodgson. Author of *Alice in Wonderland*. Logician.

J. M. Coetzee South African author and academic, winner of the 2003 Nobel Prize in Literature. BA in Mathematics at the University of Cape Town in 1961. Also BA in English, Cape Town, 1960.

Alberto Fujimori President of Peru, 1990–2000. Holds a master's degree in mathematics from the University of Wisconsin-Milwaukee.

Art Garfunkel Singer. Master's in mathematics from Columbia University. Started on his PhD, but stopped to pursue a career in music.

Philip Glass Modern composer, 'minimalist' (now 'post-minimalist') in style. Accelerated college programme in Mathematics and Philosophy, University of Chicago, at the age of fifteen.

Teri Hatcher Actor. Played Lois Lane in *The New Adventures of Superman* and also starred in *Desperate Housewives*. Mathematics and engineering major at DeAnza Junior College.

Edmund Husserl Philosopher. Mathematics PhD from Vienna in 1883.

Michael Jordan Basketball player. Started as a mathematics student at university but changed subject after his second year.

Theodore Kaczynski PhD in mathematics from the University of Michigan. Retreated to the Montana foothills and became the notorious 'Unabomber.' Sentenced to life imprisonment, with no possibility of parole, for murder.

John Maynard Keynes Economist. MA and 12th Wrangler in mathematics, Cambridge University.

Carole King Prolific pop songwriter of the 1960s, later also became a singer. Dropped out after one year of a mathematics degree to develop her musical career.

Emanuel Lasker Chess grandmaster, world chess champion 1894–1921. Mathematics professor at Heidelberg University.

J. P. Morgan Banking, steel and railroad magnate. He was so good at mathematics that the faculty of Göttingen University tried to persuade him to become a professional.

Larry Niven Author of *Ringworld* and numerous other science fiction bestsellers. Majored in mathematics.

Alexander Solzhenitsyn Winner of the 1970 Nobel Prize in literature. Author of *The Gulag Archipelago* and other influential literary works. Degree in mathematics and physics from the University of Rostov.

Bram Stoker Author of *Dracula*. Mathematics degree from Trinity College, Dublin.

Leon Trotsky Revolutionary. Studied mathematics at Odessa in 1897. Mathematical career terminated by imprisonment in Siberia.

Eamon de Valera Prime Minister and later President of the Republic of Ireland. Taught mathematics at university before Irish independence.

Carol Vorderman Highly numerate co-presenter of television series *Countdown*. Actually studied Engineering, so strictly speaking does not belong in this list.

Virginia Wade Tennis player, winner of the 1977 Wimbledon ladies' singles title. Degree in mathematics and physics from Sussex University.

Ludwig Wittgenstein Philosopher. Studied mathematical logic with Bertrand Russell.

Sir Christopher Wren Architect, in particular of St Paul's Cathedral. Science and mathematics at Wadham College, Oxford.

A Puzzling Dissection

The area can't change when the pieces are reassembled in a different way. When we form the rectangle, the pieces don't quite fit, and a long, thin parallelogram is missing – I've exaggerated the effect to show you what I mean.

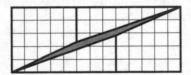

Why the area isn't 65.

In fact, if we calculate the slopes of the slanting lines, the top-left line has a slope of $2/5 = 0.4$ and the top-right line has a slope of $3/8 = 0.375$. These are different, and the first is slightly larger, so the top-left line is slightly steeper than the top-right one. In particular, they are not two pieces of the same straight line.

The key lengths in this puzzle are 5, 8 and 13 – consecutive Fibonacci numbers (page 98). You can create a similar puzzle using other sets of consecutive Fibonacci numbers.

Nothing Up My Sleeve . . .

The topological point is that because your jacket has holes,* the string is not actually linked to your body and the jacket. It just looks

---

* Armholes, not moth holes.

that way. To see that the string is not linked to your body or the jacket, imagine shrinking your body down to the size of a walnut, so that it slides down your sleeve and into your pocket. Now you can obviously pull the loop away, because your wrist is no longer blocking the gap between sleeve and pocket. However, this method is impractical, so we need a substitute.

Here's how.

Begin by pulling the end of the loop up the outside of your arm inside the jacket sleeve, as shown by the arrow in the diagram on the left. Pull out a loop at the top and draw it over your head to reach the position shown in the right-hand diagram. Then pull the loop down the outside of the other arm, inside the sleeve, as shown by the arrow in the right-hand diagram. Pull it over your hand and then back up through the sleeve. Now take hold of the string where it passes in front of your head and push it down inside the front of the jacket. The string pulls through the jacket armholes, and after a few wiggles it drops down around your ankles and you can step out of it.

Nothing Down My Leg ...

After the moves that solve the previous problem, the string ends up looped around your waist, and it is still looped around your arm. So you now follow a similar sequence of moves again, with the trousers instead of the jacket: pass a loop down the trouser leg on the side opposite the pocket with the hand in it, over the foot, back up the trouser leg – and finally remove the string down the other trouser leg. All of this is enormously undignified, and thus highly amusing to spectators. Topology can be fun.

Two Perpendiculars

Neither Euclidean theorem is wrong. Mine is.

The mistake is the assumption that P and Q are different points. In fact, P and Q *coincide* – this follows from Euclid's two theorems, and is highly plausible if you draw an accurate picture.

May Husband and Ay ...

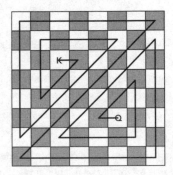

A 15-move
solution.

The smallest number of moves is 15. The path shown, and its reflection about the diagonal, are the only solutions. (Remember – each square is visited *exactly once*; that is, the path cannot cross itself.)

What Day is It?

Today is Saturday. (As I told you right at the start, the conversation took place yesterday.) Darren's answers imply that the day of the conversation is exactly one of Friday, Monday or Thursday. Delia's imply that it is Saturday, Sunday or Friday. The only common day is Friday. So *when the conversation took place*, it was Friday.

Logical or Not?

The logic is wrong. If the weather is bad, then pigs don't fly. As a consequence, we don't know whether they have wings. So we don't know whether to carry an umbrella.

It may seem strange that a deduction can be illogical when – as here – the conclusion is entirely sensible. Actually, this is very common. For example:

$$2 + 2 = 22 = 2 \times 2 = 4$$

is nonsense as far as logic goes, but it gives the right answer. All mathematicians know that you can give false proofs of correct statements. What you *can't* do – if mathematics is logically consistent, as we fervently hope – is give correct proofs of false statements.

A Question of Breeding

We are told that Catt breeds pigs.

Hamster does not breed pigs, hamsters, dogs or zebras. So he breeds cats.

Now, Dogge breeds either hamsters or zebras; Pigge breeds dogs hamsters or zebras; Zebra breeds either dogs or hamsters. Since the namesake of Zebra's animals breeds hamsters, Zebra must breed dogs. Therefore Dogge breeds hamsters, so Pigge breeds zebras.

Fair Shares

Here's Steinhaus's method. Let the three people be Arthur, Belinda and Charlie.

(1) Arthur cuts the cake into three pieces (which he thinks are all fair, hence subjectively equal).

(2) Belinda must either
- *pass* (if she thinks that at least two pieces are fair) or
- *label* two pieces (which she thinks are unfair) as being 'bad'.

(3) If Belinda passed, then Charlie chooses a piece (which he thinks is fair). Then Belinda chooses a piece (which she thinks is fair). Finally, Arthur takes the last piece.

(4) If Belinda labelled two pieces as 'bad', then Charlie is offered the same options as Belinda – pass, or label two pieces 'bad'. He takes no notice of Belinda's labels.

(5) If Charlie did nothing in step 4, then the players choose pieces in the order Belinda, Charlie, Arthur (using the same strategy as in step 3.)

(6) Otherwise, both Belinda and Charlie have labelled two pieces as 'bad'. There must be at least one piece that they *both* consider 'bad'. Arthur takes that one. (He thinks that all the pieces are fair, so he can't complain.)

(7) The other two pieces are reassembled into a heap. (Charlie and Belinda both think that the result is at least two-thirds of the cake.) Now Charlie and Belinda play I-cut-you-choose on the heap, to share what's left between themselves (thereby getting what they each judge to be a fair share).

The Sixth Deadly Sin

In the early 1960s John Selfridge and John Horton Conway independently found an envy-free method of cake division for three players. It goes like this:

(1) Arthur cuts the cake into three pieces, which he considers to be 'fair' – of equal value *to him*.

(2) Belinda must either
- *pass* (if she thinks that two or more pieces are tied for largest) or
- *trim* (the largest) piece (to make the two the same). Any trimmings are called leftovers and set aside.

(3) Charlie, Belinda and Arthur, in that order, choose a piece (one they think is largest or equal largest). If Belinda did not pass in step 2, she must choose the trimmed piece unless Charlie chose it first.

At this stage, the part of the cake other than the leftovers has been divided into three pieces in an envy-free manner—a 'partial envy-free allocation'.

(4) If Belinda passed at step 2, there are no leftovers and we are

done. If not, either Belinda or Charlie took the trimmed piece. Call this person the 'non-cutter', and the other one of the two the 'cutter'. The cutter divides the leftovers into three pieces (that he/she considers equal).

Arthur has an 'irrevocable advantage' over the non-cutter, in the following sense. The non-cutter received the trimmed piece, and even if he/she gets all the leftovers, Arthur still thinks that he/she has no more than a fair share, because he thought that the original pieces were all fair. So *however* the leftovers are now divided, Arthur will not envy the non-cutter.

(5) The three pieces of leftovers are chosen by the players in the order non-cutter, Arthur, cutter. (Each chooses the largest piece, or one of the equal largest, among those available.)

The non-cutter chooses from the leftovers first, so has no reason to be envious. Arthur does not envy the non-cutter because of his irrevocable advantage; he does not envy the cutter because he chooses before he/she does. The cutter can't envy anybody since he/she was the one who divided the leftovers.

Recently, Steven Brams, Alan Taylor and others have found very complicated envy-free methods for any number of people.

When it comes to sharing cakes, avoiding the second deadly sin* is more tricky, in my experience.

Weird Arithmetic

The *result* is correct, though as teacher said, you should cancel 9 from the top and the bottom to simplify it to $\frac{2}{5}$. But Henry's presumed *method* leaves a lot to be desired.

For instance,

$$\frac{3}{4} \times \frac{8}{5} = \frac{38}{45}$$

is wrong.

* Gluttony.

So when does his method work? An easy way to find one more
solution is to turn Henry's upside down:

$$\frac{4}{1} \times \frac{5}{8} = \frac{45}{18}$$

But there are other solutions. With the stated limits on the number of
digits, we are trying to solve the equation

$$\frac{a}{b} \times \frac{c}{d} = \frac{10a+c}{10b+d}$$

which boils down to

$$ac(10b+d) = bd(10a+c)$$

where a, b, c and d can each be any digit from 1 to 9 inclusive.

There are 81 trivial solutions where $a = b$ and $c = d$. Aside from
these, there are 14 solutions, where $(a, b, c, d) = (1, 2, 5, 4)$, $(1, 4, 8, 5)$,
$(1, 6, 4, 3)$, $(1, 6, 6, 4)$, $(1, 9, 9, 5)$, $(2, 1, 4, 5)$, $(2, 6, 6, 5)$, $(4, 1, 5, 8)$,
$(4, 9, 9, 8)$, $(6, 1, 3, 4)$, $(6, 1, 4, 6)$, $(6, 2, 5, 6)$, $(9, 1, 5, 9)$ and $(9, 4, 8, 9)$.
These form seven pairs (a, b, c, d) and (b, a, d, c), corresponding to
turning the fractions upside down.

How Deep is the Well?

The depth of the well is

$$s = \tfrac{1}{2}gt^2 = \tfrac{1}{2}10(6)^2 = 180 \text{ metres} = 590 \text{ feet}$$

which agrees very well with what the *Time Team* measured (about
550 feet) when you take into account the difficulty of timing the fall
by hand. A more accurate figure for g is 9.8 m s^{-2}, leading to a depth of
176 metres or 577 feet. Presumably the exact time was slightly less
than 6 seconds.

Yes, the well really *was* that deep. How did they dig it, so long
ago? The mind boggles.

McMahon's Squares

The 24 tiles can be assembled as shown. There are 17 other solutions,
plus rotations and reflections.

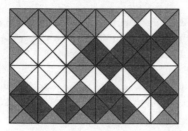

One of the 18 basically different solutions.

One feature of the tiles helps us work out how to assemble them. Choose a border colour, say grey. There are four tiles that have two blue triangles opposite each other and no other blue triangles. Their remaining triangles are grey/grey, black/black, white/white and black/white. The only way to fit these tiles in is to stack four of them together across the narrow width of the rectangle:

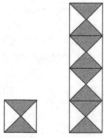

The four tiles like the left-hand one have to stack together. The white triangles can be any combination of black and white.

There are still lots of ways to proceed, but this observation helps to restrict the possibilities. There are 18 basically different solutions, which lead to 216 solutions altogether by swapping colours, rotating the picture or reflecting it. Note the stack in the third column of the sample solution above.

Archimedes, You Old Fraud!

Let's say that Archimedes can exert a force sufficient to lift his own weight, call it 100 kg. The mass of the Earth is about 6×10^{24} kg. To keep the analysis simple, suppose that the pivot is 1 metre from the Earth. Then the law of the lever tells us that distance from the pivot to Archimedes is 6×10^{22} metres, and his lever is $1 + 6 \times 10^{22}$ metres long – about 1.6 million light years, or about two-thirds of the way to the Andromeda Galaxy. If Archimedes now moves his end of the lever one metre, the Earth moves $1/(6 \times 10^{22}) = 1.66 \times 10^{-23}$ metres.

Now, a proton has a diameter of 10^{-15} metres—

Yes, but it still *moves*, dammit!

True. But suppose that instead of using this huge and improbable apparatus, Archimedes stands on the surface of the Earth and *jumps*. For every metre he goes up, the Earth moves 1.66×10^{-23} metres down (action/reaction). Jumping has exactly the same effect as his hypothetical lever. So the place to stand is on the Earth – but you don't stand *still*.

The Missing Symbol

Well, the symbols $+, -, \times, and \div$ don't work, because $4 + 5$ and 4×5 are too big, and $4 - 5$ and $4 \div 5$ are too small. Neither does the square root sign $\sqrt{}$, because $4\sqrt{5} = 8.94$ and that's too big as well.

Give up? How about the decimal point, 4.5?

Where There's a Wall, There's a Way

How to make the wall.

Rotation and reflection yield three other solutions. The component shapes are called *tetrahexes*.

Connecting Utilities

No, you can't. As stated – and without 'cooking' the puzzle by, say, working on a surface that isn't a plane, passing cables through a house, whatever – the puzzle has no solution.

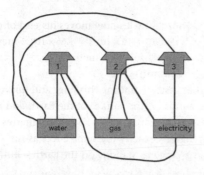

A cheat that works by passing a cable through a house.

If you experiment, you'll soon become convinced that it's impossible – but mathematicians require proof. To find one, we first connect things up without worrying about crossings, like this:

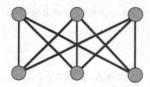

Can this be redrawn without crossings?

While we're at it, I've replaced the buildings by dots.

Now, suppose that we could redraw this picture, keeping all the connections at the dots, to eliminate the crossings. Then the lines would form a kind of map in the plane. This map would have $E = 9$ edges (the nine connections) and $V = 6$ vertices (the six dots). Euler's formula for maps (page 177) tells us that if F is the number of faces, then

$$F - E + V = 2$$

so $F - 9 + 6 = 2$ and $F = 5$. One of these faces is infinitely large and forms the outside of the whole diagram.

Now we count the edges in another way. Each face has a boundary formed by a loop of edges. You can check that the possible loops in the diagram contain either 4 or 6 distinct edges, nothing else. So there are six possibilities for the number of edges in the five faces:

```
4  4  4  4  4
4  4  4  4  6
4  4  4  6  6
4  4  6  6  6
4  6  6  6  6
6  6  6  6  6
```

which respectively total 20, 22, 24, 26, 28 and 30. But every edge forms the border between two faces, so the number of edges has to be half of one of these numbers: 10, 11, 12, 13, 14 or 15.

However, we already know that there should be 9 faces. This is a contradiction, so we can't redraw the diagram to have no crossings.

People often claim that 'You can't prove a negative.' In mathematics, you most certainly can.

Don't Get the Goat

No, there isn't. The contestant doubles their chance of success by changing their mind. But this is true *only* under the stated assumptions. For example, suppose that the host (who knows where the car is, remember) offers the contestant the opportunity to change their mind *only* when they have correctly chosen the door with the car behind it. In this extreme case, they always lose if they change their mind. At the other extreme, if he offers the contestant the opportunity to change their mind only when they have chosen a door with a goat behind it, they always win.

Fine – but what if my original assumptions are valid. The fifty–fifty argument then looks convincing, but it's wrong. The reason is that the host's procedure does not make the odds fifty–fifty.

When the contestant makes their initial choice, the probability that they have the right door is one in three. So on average and in the long run, the car is behind that particular door one time in three. Nothing that happens subsequently can change that. (Unless the television people surreptitiously move the prizes ... OK, let's assume that doesn't happen either.)

After a goat is revealed, the contestant is left with two doors. The car must be behind one of them (the host never reveals the car). One time in three that door is the one that the contestant has chosen. The other two times out of three, it must be behind the *other* door. So, if you don't change your mind, you win the car one time in three. If you do, you win it two times in three – twice the chance.

The trouble with such reasoning is that unless you've spent a lot of time learning probability theory, it's not always clear what works and what doesn't. You can experiment using dice to decide where the car goes: 1 or 2 puts it behind the first door, 3 or 4 behind the second, 5 or 6 behind the third, say. If you try this twenty or thirty times, it soon becomes clear that changing your mind really does improve the chance of success. I once got an

e-mail from some people who had been arguing about this problem in the pub, until one of them got out his laptop and programmed it to simulate a million attempts. 'Don't change your mind' succeeded on roughly 333,300 occasions. 'Do change your mind' succeeded on the remaining 666,700 occasions. It's fascinating that we live in a world where you can do this simulation in a few minutes in a pub. Nearly all of which is taken up writing the computer program – the actual sums take less than a second.

Still not convinced? Sometimes people see the light when the problem is taken to extremes. Take a normal pack of 52 playing cards, held face down. Ask a friend to pull a card from the pack, without looking at it, and lay it on the table. They win if that card is the ace of spades (car) and lose otherwise (goat). So now we have one car and 51 goats, behind 52 doors (cards). But you now pick up the remaining 51 cards, holding them so that you can see their faces but your friend can't. Now you discard 50 of those cards, none being the ace of spades. One card remains in your hand; one is on the table. Is it *really* true that each of these two cards has a fifty–fifty chance of being the ace of spades? So why were you so careful to hang on to that particular card out of the 51 you started with? Clearly you have a big advantage over your friend. They got to choose one card, without seeing its face. You had a choice of 51 cards, and you *did* see their faces. They have one chance in 52 of being right; you have 51 chances. This is a *fair* game? Pull the other one!

All Triangles are Isosceles

The mistake is the innocent assertion that X is inside the triangle. If you draw the picture accurately, it's not. And it turns out that *exactly one* of the points D and E is outside the triangle, too. In this particular case, D is not between A and C. But the other point is 'inside' the triangle—well, on its edge, but not outside it. Here E lies between B and C. This diagram makes it clear what I mean:

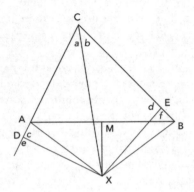

The correct picture.

Now the argument collapses. We still find that CE = CD and DA = EB (steps 5 and 9). But in step 10, CA = CD − DA, not CD + DA. However, CB is still CE + EB. So we can't conclude that lines CA and CB are equal.

Fallacies like this one explain why mathematicians are so obsessive about hidden logical assumptions in proofs.

Square Year

We are looking for squares either side of 2001. A little experiment reveals that $44^2 = 1936$, and $45^2 = 2025$. With these figures, Betty's father was born in $1936 − 44 = 1892$ (so he died in 1992), and Alfie was born in $2025 − 45 = 1980$.

To rule out any other answers: the previous possible date for Betty's father would be $43^2 − 43 = 1806$, so he would have died in 1906, making Betty well past retirement age. The next possible date for Alfie would be $46^2 − 46 = 2070$, so he wouldn't get born in 2001.

Infinite Wealth

Whatever amount you win, it will be *finite* (unless the game goes on for ever, with you always tossing tails, in which case you win an infinite amount of cash, but you have to wait infinitely long to get it). So it's silly to pay an infinite entry fee. The correct deduction is that whatever finite entry fee you pay, your expected winnings are bigger. Your chance of a big win is of course very small, but the win is so huge that it compensates you for the tiny chance of success.

But that still seems silly, and this is where the mathematicians of the time started scratching their heads (and very likely their tails too, though we don't mention such things in polite company). The main source of trouble is that the expected winnings form a *divergent series* – one with no well-defined sum – which may not make a great deal of sense.

As a practical matter, the sums involved are limited by two features that the simple mathematical model fails to take into account: the largest amount that the bank can actually pay, and the length of time available to play the game – at most one human lifetime. If the bank has only $£2^{20}$ available, for instance, which is £1,048,576, then you are justified in risking £20. If the bank has £2^{50} available, which is £1,125,899,906,842,624 – a little over a quadrillion pounds, which exceeds the annual Gross Global Product, then you are justified in risking £50.

There is a more philosophical point: how sensible the *long-term* average winnings (expectation) actually is when the 'long term' is far longer than any player can actually play for. If you are playing against a bank with £2^{50} in its coffers, it will typically take you 2^{50} attempts to make the big win that justifies you spending £50, let alone a much larger amount. Human decisions about risk are subtler than the mindless computation of long-term expectations, and the subtleties are important exactly when the gain (or loss) is very big but its probability is very small.

A related point is the relevance of long-term averages over numerous trials, if in practice you only get to play once, or just a few times. Then you have an extraordinarily small chance of a big win, and the pragmatic decision is not to throw money at something so unlikely.

On the other hand, in cases where the expectation *converges* to a finite sum, it may make more sense. Suppose that you win £n if the first toss of a head is on the nth throw. Now the expectation is

$$1 \times \frac{1}{2} + 2 \times \frac{1}{2^2} + 3 \times \frac{1}{2^3} + 4 \times \frac{1}{2^4} + \dots$$

which converges to 2. So here you should pay £2 to break even, which seems fair enough.

What Shape is a Rainbow?

The arcs are parts of circles. For a given colour, the arc concerned is very thin. All the circles involved have the same centre – which is often below the horizon. The interesting question is – why? The answer turns out to be distinctly complicated, though very elegant. Teacher was right to direct our attention to the colours, though she did miss an opportunity to do some really neat geometry.

Consider light of a single wavelength (colour), and look at a raindrop in cross-section. Raindrops are spheres, so in section we get a circle. A ray of light from the Sun hits the front of the drop, is refracted (bent through an angle) by the water, reflects off the back of the drop, and is refracted a second time as it leaves the drop and heads back roughly the way it came.

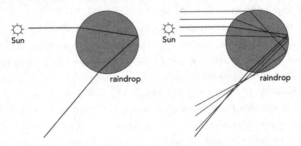

(Left) The path of one ray. (Right) Many rays.

That's what happens to one ray, but in reality there are lots of them. Rays that are very close together usually hit the same drop, but turn through slightly different angles. But there is a focusing effect, and most of the light comes back out along a single 'critical direction'. Bearing in mind the spherical geometry of the drop, the end result is that effectively each drop emits a *cone* of light of the chosen colour. The axis of the cone joins the drop to the Sun. The vertex angle of the cone is about 42°, for a raindrop, but it depends on the colour of the light.

When an observer looks at the sky, in the direction of the rain, she observes light only from those drops whose cones meet her eye. A little geometry shows that these drops themselves lie on a cone, whose tip is at her eye, and whose axis is the line joining her eye to the Sun. Again, the vertex angle is about 42°, depending on the colour of the light.

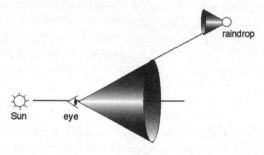

The eye receives a cone of light.

If you place a cone to your eye and sight along it, what you see is the edge of its circular base. More accurately, the *directions* of the incoming light are perceived as if the light were being emitted by the circular base. So the upshot is that the eye 'sees' a circular arc. The arc is not up there in the sky: it is an illusion, caused by the directions of the incoming light rays.

Usually, the eye sees only part of this circular arc. If the Sun is high in the sky, most of the arc is below the horizon. If the Sun is low, the eye sees almost a semicircle. From an aircraft a complete circle can sometimes be seen. If the rain is nearby, the arc may appear to be in front of other parts of the landscape. The arc is often partial – you see the returning light only when there's rain in that direction.

Because different colours of light lead to different vertex angles for the cone, each colour appears on a slightly different arc, but they all have the same centre. So we see 'parallel' arcs of colours.

Sometimes you can see a second rainbow, outside the first. This is formed in a similar way, but the light bounces more times before coming back out of the drop. The vertex angle of the cone is different, and the colours are in reverse order. The sky is brighter inside the main rainbow, very dark between this and the 'secondary' rainbow, and medium-dark outside that. Again, all this can be explained in terms of the geometry of the light rays. René Descartes did that in 1637.

A really informative website is **en.wikipedia.org/wiki/Rainbow**

Alien Abduction

Each alien is going after the pig that initially is nearest to it. If they chase the *other* pig, they will soon catch it.

Why? The way to catch a pig is to drive it into a corner. If the position looks like the next picture, and it is the *pig's* turn to move, it will be abducted. However, if it is the *alien's* turn to move, the pig can escape.

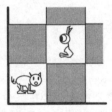

How to catch your pig
– provided the pig has to move.

Which of these happens depends on the *parity* (odd or even) of the distance (in moves) from alien to pig. If the pig is an even number of moves away – as it is if each alien goes for the pig it is initially facing – then the pig always escapes. If it is odd – as it is if the aliens switch pigs – then the pig can be driven into a corner and abducted.

Disproof of the Riemann Hypothesis

The argument is logically correct. However, it doesn't disprove the Riemann Hypothesis! The information given is contradictory: it implies that an elephant has won *Mastermind*, and also that it has not. We can now prove the Riemann Hypothesis false by contradiction:

(1) Assume, on the contrary, that the Riemann Hypothesis is true.

(2) Then an elephant has won *Mastermind*.

(3) But an elephant has not won *Mastermind*.

(4) This is a contradiction. So our assumption that the Riemann Hypothesis is true is wrong.

(5) Therefore the Riemann Hypothesis is false.

The same argument proves that the Riemann Hypothesis is true, of course.

Murder in the Park

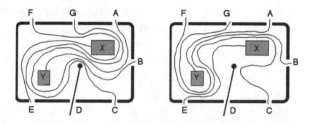

The two possible topological types of path.

Topologically speaking, there are just two cases to consider. Either the butler went to the north of Y on his way to X (left-hand diagram) or he went to the south (right-hand diagram). The gamekeeper must then have gone to the south (respectively north) of X on his way to Y.

The tracks of the youth and the grocer's wife must then be as shown, perhaps with additional wiggles. In the first case, the grocer's wife's path from C to F cuts off the youth's path from the part of the park that contains the body. In fact, only she and the butler could have approached the place where Hastings's body lay. The same is true in the second case. Since the butler has a confirmed alibi, the murderer must have been the grocer's wife.

The Cube of Cheese

The corners of the hexagon lie on various midpoints of sides of the cube, like this:

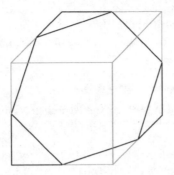

Hexagonal slice of a cube.

Drawing an Ellipse – and More?

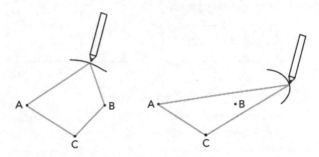

The pencil draws arcs of various ellipses.

When the pencil is in the position shown on the left, the length of string AC+CB is constant, so the pencil moves as if you had looped a shorter string round A and B. Therefore it draws an arc of an ellipse with foci A and B. When it moves to a position like the one on the right, it draws an arc of an ellipse with foci A and C. The complete curve therefore consists of six arcs of ellipses, joined together. Since basically this isn't new, mathematicians aren't (terribly) interested.

The Milk Crate Problem

The milkman is correct for 1, 4, 9, 16, 25 and 36 bottles, but wrong for 49 and any larger square number.

If you think about this the right way, it's obvious that when the number of bottles gets sufficiently large, the square lattice packing can't possibly be the best. (What *is* the best is horribly difficult to work out, and nobody knows.) The square lattice must fail for a large number of bottles, because a hexagonal lattice packs bottles more closely than a square one. When there aren't too many bottles, 'edge effects' near the walls of the crate stop you exploiting this fact to make the crate smaller, but as the numbers go up the edge effects become negligible.

It so happens that the break-even point is close to 49 bottles. And it has been proved that 49 bottles of unit diameter can fit inside a square whose side is very slightly less than 7 units. The difference is too small to be seen by the naked eye, but you can easily see big regions of hexagonally packed circles.

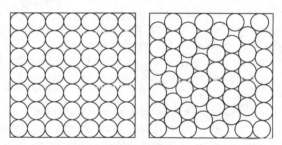

(Left) 49 unit bottles in a 7 × 7 square. (Right) How to fit the same bottles into a slightly smaller square.

Incidentally, this example shows that a *rigid* packing – one in which no single circle can move – need not be the closest packing possible. The square lattice is rigid for any square number of bottles inside a tightly fitting square crate. Or, indeed, on the infinite plane.

Road Network

The shortest road network introduces two new junctions and makes the roads meet there at *exactly* 120° to each other. The same layout rotated by 90° is the only other option. The total length here is $100(1 + \sqrt{3}) = 273$ km, roughly:

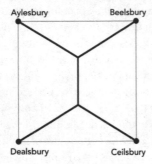

The shortest network.

Tautoverbs

- No news is no news. (Experts consider this the smallest but most perfectly formed tautoverb, a kind of tautohaiku.)
- The bigger they are, the bigger they are.
- Nothing ventured, nothing lost.
- Too many cooks cook too much.
- You cannot have your cake and eat it too, *unless you do them in that order*. What's difficult is to eat your cake and have it too.
- A watched pot never boils *over*. (Unless it's custard.) The time taken for a liquid to boil is not influenced by the presence of an observer, except in certain esoteric forms of quantum field theory. It merely seems longer for psychological reasons. Do not be deceived.
- If pigs had wings, the laws of aerodynamics would still stop them getting off the ground. I mean, let's be *sensible*. The porcithopter is not technologically feasible.

Scrabble Oddity

TWELVE $= 1 + 4 + 1 + 1 + 4 + 1$

Dragon Curve

Dragon curves can be made by repeatedly folding a strip of paper in half – always folding the same way – and then opening it out to make all folds into right angles.

Making dragon curves by paper-folding.

These curves determine a fractal (page 189). In fact, the infinite limit is a space-filling curve (page 83), but the region it fills has a complicated, dragon-like shape. The sequence of right (R) and left (L) turns in the curve goes like this:

Step 1 R

Step 2 R **R** L

Step 3 R R L **R** R L L

Step 4 R R L R R L L **R** R R L L R L L

In fact, there is a simple pattern: each sequence is formed from the previous one by placing an extra R at the end, followed by the reverse of the previous sequence with R's and L's swapped. I've marked the extra R in the middle in bold.

The dragon curve was discovered by John Heighway, Bruce Banks and William Harter – all physicists at NASA – and was mentioned in Martin Gardner's Mathematical Games' column in *Scientific American* in 1967. It has lots of intriguing features – see en.wikipedia.org/wiki/Dragon_curve

Counterflip

Assume that there is an odd number of black counters—so in particular, there exists at least one of them. As play progresses, counters that are removed create gaps, breaking the row of counters into connected pieces, which I'll call *chains*. We start with one chain.

I claim that any chain with an odd number of black counters can be removed. Here's a method that always works. (The sample game doesn't always follow it, so other methods work too.)

Starting from one end of the chain, find the first black counter. I claim that if you flip that counter, then there are three possibilities:

(1) The chain originally consisted of one isolated black counter, and when you flip it it is removed, with no effect on any other counters.

(2) You now have a single shorter chain having an odd number of black counters.

(3) You now have two shorter chains, each having an odd number of black counters.

If this claim is true, then you can repeat the same procedure on the shorter chains. The number of chains may grow, but they get shorter at each step. Eventually they all become so short that we reach case 1 and they can be removed entirely.

The claim is proved by seeing what happens in three cases of a single chain, which exhaust the possibilities:

(1) The chain concerned consists of a single black counter. It has no neighbours, so when it is flipped it disappears.

(2) The chain has a black counter at one end. Flipping the end counter results in a shorter chain which has an odd number of black counters.

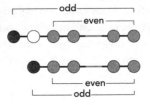

Grey counters may be either black or white. The black counter on the end disappears, and its neighbour (here shown white) changes colour. The overall change in the number of black counters is either 0 or 2, so an odd number of black counters remains.

(3) The chain has white counters at both ends. Flipping the first black counter from one end (it doesn't matter which) results in

two shorter chains. One has a single black counter (which is odd) and the other has an odd number of black counters.

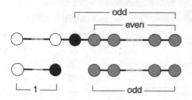

The first black counter (from the left) disappears, and its neighbours (one white, one grey – i.e. black or white) change colour. Two chains are created; one has one black counter, the other has an odd number of black counters.

It does matter which black counter you flip. For instance, if the chain has at least four counters, with three black counters next to one another and the rest white, then it is a mistake to flip the middle black counter. If you do, you get at least one chain containing no black counters at all, so this chain cannot be removed.

Oops ...

To complete the analysis, here's why the puzzle can't be solved if the initial number of black counters is even:

(1) If there are no black counters (zero is even!) you can't get started.

(2) If the initial number of black counters is even (and non-zero), then whichever black counter you remove, at least one shorter chain is created that also has an even number of black counters. Repeating this process eventually leads to a chain with no black counters but at least one white one. This chain cannot be removed since there is no place to start.

Spherical Sliced Bread

All slices have exactly the same amount of crust.

At first sight this seems unlikely, but slices near the top and bottom are more slanted than those near the middle, so they have more crust than you might think. It turns out that the slope exactly compensates for the smaller size of the slices.

In fact, the great Greek mathematician Archimedes discovered that the surface area of a slice of a sphere is equal to that of the corresponding slice of a cylinder into which the sphere fits. It is obvious that parallel slices of a cylindrical loaf, of equal thickness, all have the same amount of crust ... since they are all the same shape and size.

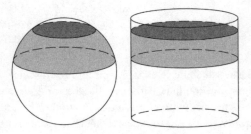

The surface area of the spherical band (pale blue) is the same as that of the corresponding band on a tightly fitting cylinder.

Mathematical Theology

I asked you to start from $2 + 2 = 5$ and prove that $1 = 1$ and also that $1 = 42$. There are lots of valid answers (infinitely many, in fact). Here are two that work:

- Since $2 + 2 = 4$, we deduce that $4 = 5$. Double both sides to get $8 = 10$. Subtract 9 from each side to get $-1 = 1$. Square both sides to get $1 = 1$.
- Since $2 + 2 = 4$, we deduce that $4 = 5$. Subtract 4 from each side to get $0 = 1$. Multiply both sides by 41 to get $0 = 41$. Add 1 to each side to get $1 = 42$.